ASTM INTERNATIONAL

SELECTED TECHNICAL PAPERS
STP1635

Editors: Daniel J. Lemieux and Jennifer Keegan

Building Science and the Physics of Building Enclosure Performance: 2nd Volume

ASTM STOCK #STP1635
DOI: 10.1520/STP1635-EB

ASTM International, 100 Barr Harbor Drive, PO Box C700, West Conshohocken, PA 19428-2959
Printed in the U.S.A.

Library of Congress Cataloging-in-Publication Data
Names: Keegan, Jennifer, editor. | Lemieux, Daniel J., editor.
Title: Building science and the physics of building enclosure performance /
    editors, Jennifer Keegan, Daniel J. Lemieux.
Description: West Conshohocken : ASTM International, 2020. | Series:
    Selected technical papers STP1617 | Includes bibliographical references.
    | Summary: "This compilation of Selected Technical Papers, STP1617,
    Building Science and the Physics of Building Enclosure Performance,
    contains peer-reviewed papers that were presented at symposiums held
    October 21-22, 2018, and December 2, 2018, in Washington, DC, USA. The
    symposiums were sponsored by ASTM International Committee E06 on
    Performance of Buildings and Committee D08 on Roofing and
    Waterproofing"- Provided by publisher.
Identifiers: LCCN 2020007959 (print) | LCCN 2020007960 (ebook) | ISBN
    9780803176805 (paperback) | ISBN 9780803176812 (pdf)
Subjects: LCSH: Buildings–Thermal properties. | Dampness in buildings.
Classification: LCC TH6025 .B827 2020 (print) | LCC TH6025 (ebook) | DDC
    720/.472-dc23
LC record available at https://lccn.loc.gov/2020007959
LC ebook record available at https://lccn.loc.gov/2020007960

ISBN: 978-0-8031-7716-1
ISBN-EB: 978-0-8031-7717-8

Copyright © 2022 ASTM INTERNATIONAL, West Conshohocken, PA. All rights reserved. This material may not be reproduced or copied, in whole or in part, in any printed, mechanical, electronic, film, or other distribution and storage media, without the written consent of the publisher.

Photocopy Rights

Authorization to photocopy items for internal, personal, or educational classroom use, or the internal, personal, or educational classroom use of specific clients, is granted by ASTM International provided that the appropriate fee is paid to the Copyright Clearance Center, 222 Rosewood Drive, Danvers, MA 01923, Tel: (978) 646-2600; http://www.copyright.com/

ASTM International is not responsible, as a body, for the statements and opinions expressed in this publication. ASTM International does not endorse any products represented in this publication.

Peer Review Policy

Each paper published in this volume was evaluated by two peer reviewers and at least one editor. The authors addressed all of the reviewers' comments to the satisfaction of both the technical editor(s) and the ASTM International Committee on Publications.

The quality of the papers in this publication reflects not only the obvious efforts of the authors and the technical editor(s), but also the work of the peer reviewers. In keeping with long-standing publication practices, ASTM International maintains the anonymity of the peer reviewers. The ASTM International Committee on Publications acknowledges with appreciation their dedication and contribution of time and effort on behalf of ASTM International.

Citation of Papers

When citing papers from this publication, the appropriate citation includes the paper authors, "paper title," in *STP title*, book editor(s) (West Conshohocken, PA: ASTM International, year), page range, paper doi, listed in the footnote of the paper. A citation is provided on page one of each paper.

Printed in Batavia, NY
May, 2022

Foreword

THIS COMPILATION OF Selected Technical Papers, STP1635, *Building Science and the Physics of Building Enclosure Performance: 2nd Volume*, contains peer-reviewed papers that were presented at symposiums held April 24, 2022, and June 12, 2022, in Seattle, Washington, USA. The symposiums were sponsored by ASTM International Committee D08 on Roofing and Waterproofing, Committee E06 on Performance of Buildings, and Committee C16 on Thermal Insulation.

Symposium Chairs and STP Editors:

Jennifer Keegan
*GAF Materials Corporation
Parsippany, NJ, USA*

Daniel J. Lemieux
*Wiss, Janney, Elstner Associates, Inc.
Washington, DC, USA*

Contents

Overview	vii
Calibrated Energy Modeling Importance—A Sensitivity Analysis of Building Enclosure Performance Factors John M. Civitillo Jr. and Victoria T. Civitillo	1
Coordination of ASTM Standards for Thermal Insulation Materials Installed in Exterior Cavity Wall Applications Andrew W. Wagner and Jodi M. Knorowski	29
Pillow Talk: ETFE Roofing Design and Quality-Control Testing Martina T. Driscoll, Andrea DelGiudice, and Adrienne Larson	53
An Investigation of Stucco Wall Assembly Performance by Hygrothermal Simulation Theresa A. Weston	71
Moisture Load for a Vinyl-Clad Wall Assembly for Selected Canadian Cities Zhe Xiao, Michael A. Lacasse, Maurice Defo, and Elena Dragomirescu	90
Design and Detailing Air and Weather Barrier Transitions to Fenestrations Derek J. Ziese	101
Considerations for Unitized Building Enclosure Systems: Selecting, Designing, Installing, and Testing Unitized Building Enclosure Systems Emily R. Hopps and Anna M. Burhoe	133
Three-Dimensional Thermal and Condensation Risk Analysis of Cantilevered Curtain Wall Elements Andrew A. Dunlap and Ryan Asava	144

Managing Interfaces in Complex Systems Andrea Wagner Watts and William Ranson	174
Low-Slope Roofing Installations and Third-Party Observations: A Critical Review of Noncompliance Management Benjamin Meyer, Keith Nelson, and Kristin Westover	195
Evaluation of Existing Steep Sloped Roof Assemblies for Changes to Thermal, Moisture, and Ventilation Performance: A New Joint E06/D08 Standard in Action Carly May Wagner and Rex A. Cyphers	218
Experimental Thermal Resistance Comparison of a Reflective Insulation in Vertical and Horizontal Furred Airspace Orientations Using a Guarded Hot Box Travis V. Moore, Mehdi Ghobadi, Alex T. Hayes, and Josip Cingel	238
Moisture Management in Wood Volumetric Modular Construction Kyle H. Jang, Elyse A. Henderson, and Graham Finch	252
Evaluation of Chemical Permeation Performance of Polymeric Below-Grade Waterproofing Membranes Xia Cao and Nicholas D. Anderson	269
Evaluation Methods, Hazard Risk Assessment and Mitigation of Building Enclosure Design Elements that are Prone to Ice and Snow Collection and Release Michael Carter	283
Look Out Below! Addressing Falling Snow and Ice Hazards on Building Facades Sean M. O'Brien and Amarantha Z. Quintana-Morales	298

Overview

The first ASTM International Joint Symposium on Building Science and the Physics of Building Enclosure Performance was held in Washington, DC. It was a two-part symposium that began in October 2018 with Committee E06 on Performance of Buildings and concluded in December of that year with Committee D08 on Roofing and Waterproofing.

Originally conceived as a recurring series that would be hosted by ASTM International in collaboration with similarly aligned standards development and professional interest organizations, the first symposium brought together thought leaders from across North America, the UK, Europe, and Asia to address the challenges that we continue to face with building enclosure and whole building performance.

The overwhelming success of the first symposium and the concerns that it raised relative to building performance and the contribution of our built environment to climate change led immediately to planning for the second symposium in this series, originally scheduled for 2020 in Toronto, Ontario, Canada. Sadly, for so many of us and for our co-hosts—ASTM Committee C16 on Thermal Insulation, National Research Council Canada, and the International Institute of Building Enclosure Consultants—those plans were interrupted by COVID-19 and what quickly became a global pandemic that would interrupt our lives and, in a variety of different ways, perhaps forever change the ways in which we live and work.

Today, as we gather again for the first time in over two years, we are very pleased to welcome you to the city of Seattle, home of the Second Symposium on Building Science and the Physics of Building Enclosure Performance. We recognize the difficult and often deeply personal decisions that so many of you have very likely had to make over the past two years and we truly appreciate your decision to join us for this event.

We would also like to take this opportunity to offer our sincere thanks to Frank McConnell and Steve Mawn, our respective ASTM Staff Managers for Committees D08 and E06, for their tireless patience and support. We share a common goal to encourage a more cooperative, cross-discipline and fully transparent approach to standards development at ASTM. Frank, Steve, and the ASTM Board of Directors are firm believers in that effort. This symposium and the symposia that will follow are in no small part a direct result of their insight, wisdom, and learned guidance.

Many thanks also to Kelly Dennison, Alyssa Conaway, Sara Welliver, and all of the truly dedicated, professional staff at ASTM for making this second joint symposium possible during what has been a very challenging two years for all of us. Teamwork and a considerable amount of patience—often beyond our view and behind the curtain—came together to make it all look so easy.

And finally, a very special thanks to our speakers, authors, presenters, technical reviewers, and invited and registered guests for donating your time, interest, passion and expertise. As we have said in the past, for both Part I and Part II of this symposium, we are your students—humbled by your knowledge and grateful for your continued contributions to our industry and to the advancement of our profession. It has been a privilege to serve each of you and to serve ASTM as co-chairs of this symposium.

Symposium Chairs and STP Editors:

Jennifer Keegan, AAIA
Director
Building and Roofing Science
GAF Materials Corporation

Daniel J. Lemieux, FAIA, NCARB, RIBA, MRICS
Director and Principal
Wiss, Janney, Elstner Associates, Inc. (USA)
Wiss, Janney, Elstner Limited (UK)

John M. Civitillo Jr.[1] and Victoria T. Civitillo[2]

Calibrated Energy Modeling Importance—A Sensitivity Analysis of Building Enclosure Performance Factors

Citation

J. M. Civitillo Jr. and V. T. Civitillo, "Calibrated Energy Modeling Importance—A Sensitivity Analysis of Building Enclosure Performance Factors," in *Building Science and the Physics of Building Enclosure Performance: 2nd Volume*, ed. D. J. Lemieux and J. Keegan (West Conshohocken, PA: ASTM International, 2022), 1–28. http://doi.org/10.1520/STP163520210030[3]

ABSTRACT

The green building movement has motivated many governing bodies to pursue regulation of building energy efficiency. In 2019, New York City introduced unprecedented legislation targeting the operational energy consumption of the existing building stock. The intent of Local Law 97 is to impose increasingly rigorous limitations on greenhouse gas emissions from buildings 25,000 sq. ft. or greater. Based on current building utility data, approximately 75% of buildings subject to this legislation will not comply with limits set for 2030. Consequently, this legislation is expected to lead building owners to pursue energy modeling to predict future energy usage and evaluate various energy conservation measures. Energy models for existing buildings must be calibrated on the basis of current HVAC equipment, operational schedules, occupancy rates, lighting loads, process loads, facade construction, climatic data, and historical utility consumption. When dealing with existing buildings, documentation is often not available regarding the as-built conditions of the building systems or enclosure

---

Manuscript received March 15, 2021; accepted for publication June 17, 2021.
[1] WDP Consulting Engineers, P.C., 202 W. 40th St., Ste. 301, New York, NY 10018, USA http://orcid.org/0000-0002-9724-7477
[2] P&A Consulting Engineers PLLC, 19 W. 21st St., Ste. 705, New York, NY 10010, USA http://orcid.org/0000-0002-5912-4337
[3] ASTM Second Symposium on *Building Science and the Physics of Building Enclosure Performance* on April 24–25, 2022, and June 12, 2022 in Seattle, WA, USA.

Copyright © 2022 by ASTM International, 100 Barr Harbor Drive, PO Box C700, West Conshohocken, PA 19428-2959.

ASTM International is not responsible, as a body, for the statements and opinions expressed in this paper. ASTM International does not endorse any products represented in this paper.

performance. To develop an energy model, assumptions must be made regarding the building enclosure, including the opaque thermal performance, air exfiltration and infiltration, fenestration performance, or the presence of thermal bridging. These assumptions can result in conservative inputs that do not reflect the built condition. The building enclosure performance has a direct impact on the operational performance of the HVAC systems and therefore the overall building energy consumption. This paper explores the impact of variations in building enclosure performance assumptions on the overall building energy usage. It provides recommendations for investigative methods to quantify the performance of the existing building enclosure to improve modeling accuracy. These methods include but are not limited to instrumentation to determine in situ R-values, blower door testing to measure air infiltration and exfiltration, and National Fenestration Rating Council simulation to define fenestration performance. Finally, this paper determines the relative importance of each variable in energy model calibration.

Keywords

calibrated energy modeling, existing building, sensitivity, building enclosure, field verification

Introduction

In April 2019, New York City issued the Climate Mobilization Act, one of the farthest-reaching sets of local laws aimed to tackle carbon emissions across buildings, vehicles, city-wide practices, and more. One of the most progressive pieces within this legislation is Local Law 97, which introduces annual carbon emissions caps for all buildings greater than 25,000 sq. ft. across the city. The primary goal of the law is to reduce buildings' greenhouse gas emissions by 40% by 2030 relative to a 2005 baseline. Although many countries and regions have codified building regulations intended to limit carbon emissions, they primarily govern the design and construction of new buildings. This legislation is one of the first to place a limit on the operational energy consumption of an existing building stock, although other localities have introduced similar standards or are expected to follow suit with legislation. The law targets buildings, which account for nearly three-quarters of the carbon emissions of the city, and it is estimated that 85% of existing buildings that are on the current skyline will remain in 2030.[1]

The annual energy consumption limitations of Local Law 97 are determined on the basis of a given building's occupancy category and floor area compared with the specific equivalent carbon emission for each fossil fuel consumed.[2] The first restrictions become effective in 2025 and are aimed to target roughly the worst performing 20% of buildings in the city.[3] These initial limitations will remain constant for subsequent years until 2030, at which point the allowable equivalent carbon emissions will be halved. Approximately three-quarters of all buildings greater than 25,000 sq. ft. in New York City are anticipated to exceed the 2030 limitations at their current consumption levels.[3] In general, it can be expected that these consumption levels will increase as heating and cooling demands become greater as the current climate change trends intensify. Property owners of buildings that do not comply with the respective carbon emissions cap are subject to pay a steep fine based on a building's equivalent carbon consumption in excess of the calculated limit. Although the specific boundaries have not been set, two subsequent reductions in the permissible equivalent carbon emission are projected for 2040 and 2050. The potential resulting fines from Local Law 97 have shaken the real estate market. Property owners, building engineers, real estate boards, consulting engineers, and architects are already rushing to determine how an existing building's performance stacks up in relation to its projected limitations.

The reality is that most existing buildings greater than 25,000 sq. ft. will need to take steps to reduce their energy consumption to comply with progressively tightening restrictions. This has created a shift in the way the architecture, engineering, and construction industries in New York City are approaching energy modeling. Until recently, the primary application for energy modeling was comparative. Energy models were created during the design phase of new construction to demonstrate a predicted improvement over a given baseline, be it a comparison against ASHRAE 90.1 or similar. These design-phase energy models are based on assumptions of construction quality, operation schedules, maintenance procedures, occupant behavior, heating and cooling set points, equipment configuration, and efficiencies. They rarely predict a building's actual energy demand with a high degree of accuracy. The annual energy consumption demonstrated in an energy model created during the design phase of building can vary from 5% to as much as 97% from the building's actual energy consumption postconstruction.[4] It has been documented that the differences in installed performance can be a result of the variability in the experience of the installer and differing levels of involvement of the design team.[5]

Looking forward, many owners will be pursuing energy modeling of existing buildings not only as an important means to understand the existing energy demands of their property but also as a tool to evaluate the costs and benefits of implementing potential energy conservation measures. The benefit of energy modeling of an existing building is that the actual energy demands are known through annual utility data, and a model can be tweaked by updating inputs so that the modeled outputs match. Although this task might sound simple in concept, the importance of being accurate in all model assumptions and tweaking the correct inputs the appropriate amount cannot be overstated. As a simplified example, if an energy modeler needs to tweak a model input to account for additional electricity consumed, he or she may adjust the occupancy schedule to find that increasing the total artificial lighting load means the energy model yields the correct energy expenditure of the building. The owner may overestimate the importance of lighting loads on the basis of this decision and decide to replace all lights with LED fixtures. But if the additional load were actually attributed to the HVAC systems and not lighting, the predicted energy savings shown in the model would misrepresent the actual savings the building would realize. To maximize accuracy, a parametric calibration of the energy model requires an in-depth understanding of how the HVAC systems, lighting systems, process loads, occupants, and building enclosure will all interact and operate within a building. The most realistic inputs are those that are based on field-verified parameters.

In some cases, it is easy to field-verify inputs. Mechanical systems are often marked by the manufacturer with the make, model, capacity, efficiency, and other pertinent operational information. Processes such as energy audits and submetering can help to classify energy distribution at a granular level. However, to truly understand space conditioning loads takes great effort. For the purposes of this study, building energy consumption can be categorized as either space conditioning loads or several other loads (domestic hot water, lighting, appliances, plug loads, etc.). Space conditioning, which is typically around 30%–40% of a commercial office building's energy use, can be considered to be made up of two parts: mechanical equipment that physically consumes energy to produce heating, cooling, and ventilation, and the building enclosure, whose physical properties define the boundaries between interior and exterior spaces (walls, windows, roofs, foundation). The building enclosure directly influences the loads that the mechanical equipment is required to meet. Oftentimes, the building enclosure assemblies and materials are not well documented, making it difficult to accurately input the numerous associated parameters required by a model. Further complicating the uncertainty of the building enclosure is the fact that the constituent layers that make up the exterior wall and roof systems are typically concealed by interior finishes and exterior cladding.

Many practitioners recognize the inaccuracy inherent in energy modeling. There are inaccuracies both due to the calculation (software) itself as well as the accuracy of assumptions and inputs. Some have even suggested the use of safety factors in prediction-based models.[6] This paper does not address the accuracy of modeling software but instead focuses on improving accuracy through better assumptions and inputs. This sensitivity study explores the relative importance of various building enclosure assumptions made within an energy model (namely the air leakage, opaque wall thermal resistivity, and fenestration performance). The approach is two-fold: to understand where to prioritize efforts in field verification for model calibration and then to prioritize key components of the enclosure that can be improved in a retrofit project to reduce a building's overall energy consumption.

Methods and Results

BUILDING EXPLORED AND MODELING SOFTWARE

All quantitative results that are presented relate to a particular case study: a 50-story commercial office building located in Midtown Manhattan, the central business district of New York City. The building has a retail component on the first floor and is configured with a parking garage at the cellar and subcellar levels. The building, constructed in 1968, estimates approximately 1,932,000 sq. ft. with a curtain wall system in which the opaque facade consists of an insulated spandrel glass system. Although there is no "typical" commercial high-rise building in New York City, this building was chosen because it is comparable in size and construction type to roughly one-third of the commercial office buildings in Manhattan. The structure is roughly 645-ft. tall with slab-to-slab heights ranging from 12'-1" to 14'-11" on commercial tenant floors with a lobby and retail slab height of 30'-0". Commercial tenant floors are modeled with private offices and conference rooms occupying the perimeter spaces, whereas open office areas, pantries, print areas, restrooms, and IT rooms occupy the core space of the floorplate. The building facade is approximately 487,000 sq. ft. with a 45% window-to-wall ratio. The base building heating, ventilation, and air-conditioning units rely on a chilled water distribution system for cooling, a condenser water system for heat rejection, and a steam distribution system for heating.

This sample case is a well-documented real office building in New York City. Although several of the existing building enclosure assemblies and components are known, their assumed configuration and performance were varied individually to explore the relationship between total predicted energy consumption and these variables. The resulting relationships and calculated influence coefficients are therefore specific to this base building configuration.

TRACE® 700 was the modeling software used for this analysis.[7] This program is primarily used as an HVAC load calculation tool but has extended building energy modeling capabilities. There are several other prominent modeling programs used within the industry that would also be appropriate for use in this study. These include eQuest, EnergyPlus^{TM}, and iES-VE. At a high level, these programs differ in their ability to model certain building characteristics (e.g., HVAC systems). Less consequential differences in features include model run times, quantity of parametric runs, and software support tools. It was important to use a performance-based energy modeling tool for this analysis rather than a prescriptive-based one such as COMcheck^{TM}. Prescriptive-based software is limited in its capacity to illustrate how building characteristics such as the envelope, HVAC systems, and lighting interact and influence one another.

BUILDING ENCLOSURE PERFORMANCE FACTORS

The building enclosure (or envelope) is the physical assembly or system of assemblies that make up the environmental separation layers between the interior environment and exterior environment. The enclosure consists of a number of different systems, such as foundations, slabs on grade, exterior walls, and roofs. Each of these systems are often subdivided into multiple assemblies and components (windows, doors, multiple facade types, and cladding), each of which has distinct physical properties and affects the relationship between the interior and exterior environment differently. Although the building enclosure can be considered in relation to many environmental elements, including light and acoustics, it is typically thought of in terms of the three primary environmental controls: heat, air, and vapor. This study focuses on three associated primary assembly properties: thermal resistance, solar radiation of glazed areas, and whole building air leakage.

Baseline parameters were established for each category of building performance factor examined and were based on industry standards, guidelines, and the collective experience of the authors to approximate realistic installed building performance metrics. Once these control building performance metrics were established (**table 1**), each performance factor was varied within a reasonable extent to explore the effect on overall building energy consumption.

Opaque Assembly R-Value

R-value is the standard measure of the thermal resistance to the transmission of heat energy by conduction across a thermal boundary subject to a temperature differential. Also frequently used, the thermal conductivity (U-factor) of an assembly is the inverse of thermal resistance. For the opaque wall assemblies, two distinct baseline assemblies were modeled to represent two typical assemblies found in existing construction in New York City: a mass masonry wall and a light-gauge metal-framed construction with continuous insulation. The intent was to understand any differences between the two types of construction, particularly given the difference in thermal masses.

The metal-framed assembly, from the outside in, was assumed to be composed of an exterior air film, 3/8 in. stucco, 5/8 in. exterior sheathing, 2 1/4 in. R-7.5 continuous insulation, 3 1/2 in. R-13 cavity batt insulation (between light gauge stud framing at 16 in. on center), 5/8 in. gypsum board, and an interior air film. Together, this assembly constitutes the ASHRAE 90.1 minimum performance level for a steel-framed above-grade wall in Climate Zone 4A,[8] with a thermal conductivity of U-0.064 BTU/h/sq. ft./°F (resistivity of R-15.6 h·sq. ft.·°F/BTU). Additional models were then created to calculate the performance of the building at slightly different levels of thermal resistance.

The mass masonry wall assembly was assumed to represent a typical uninsulated three-wythe load-bearing brick masonry structure. From the outside in, it consisted of an exterior air film, 4 in. face brick, 4 in. common brick, another layer of 4 in. common brick, 3/4 in. plaster at the interior surface, and an interior air film.

TABLE 1 List of control energy model inputs

Building Enclosure Performance Factor	Value
Air leakage	0.055 cfm/sq. ft. of wall
Window U-value	0.36
Window SHGC	0.4
Opaque wall assembly U-value	0.06413 (metal-framed)

This assembly was calculated to have a U-factor of 0.31 BTU/h/sq. ft./°F (R-3.2 h·sq. ft. ·°F/BTU). A series of models was run varying the total R-value for this assembly to explore the sensitivity of the model to assembly thermal performance at this end of the spectrum.

Fenestration Performance
As with opaque wall assemblies, windows and other fenestrations also have a thermal resistance. These thermal resistance values can be more complex to calculate or measure because the overall fenestration is made up of varying amounts of glass and frame, both of which have different thermal properties. Typically, the overall fenestration U-factor is used, averaging the effect of the framing and glass over the area of the system. For this analysis, the U-factor for the existing building was assumed to be 0.36 (BTU/h·sq. ft.·°F). This value was increased and decreased during the energy model iterations to explore trends.

Windows and glazed fenestrations are distinct from opaque assemblies in that they are susceptible to permit additional heat transfer in the form of solar radiation along with visible light. Solar radiation is unique in that the direction of heat transfer is always toward the interior. Where thermal resistance attempts to limit the conduction of heat across an assembly, it will switch directions seasonally to temper the conditioned space from the exterior, whether it be warmer or cooler than the set-point temperature. The result of this effect is particularly complex in mixed climates such as New York City, such that solar radiation may be beneficial in the winter months to offset heating loads but should not permit so much radiation as to overwhelm cooling loads in the summer months. It may also be desirable to install fenestrations with different resistances to solar radiation on different elevations of the building depending on their exposure.

For the purpose of this study, the window solar heat gain coefficient (SHGC) was assumed to be constant for all glazed fenestrations. The SHGC is the ratio of solar radiation transmitted to the total incident radiation. This study explored a wide range of typical SHGCs from 0.3 to 0.6. Instead of inputting the SHGC directly, TRACE 700 relies on the window shading coefficient (SC). The SC is defined as the shading effect of the window glass type.[7] It is a direct multiplier on the solar radiation load through the glass. A value of 1.0 indicates no shading effect, whereas a value of 0 indicates a total (complete) shading effect resulting in no solar radiation load. This entry affects the glass solar load of the room. The SC at normal incidence is equivalent to the SHGC at normal incidence divided by a factor of 0.87.[7] For this analysis, normal incidence was assumed, and thus the SC modifications were made such that the values were equivalent to the SHGC divided by 0.87.

Air Leakage
Air leakage was modeled as infiltration within the TRACE 700 model. ASHRAE 90.1-2016 defines infiltration as, "the uncontrolled inward air leakage through cracks and crevices in any building element and around windows and doors of a

building caused by pressure differences across these elements due to factors such as wind, inside and outside temperature differences (stack effect), and imbalance between supply and exhaust air systems."[8]

An infiltration input in the energy model is intended to quantify the transfer of outside air into the building through the enclosure (typically through windows, leaky construction, or doors). The software allows the option to differentiate between cooling infiltration airflow and the heating infiltration airflow. For the purposes of the building model the infiltration rates were kept constant during cooling and heating conditions year-round. The infiltration rates were entered with the units CFM/sq. ft. of wall. Per ASHRAE 90.1 Appendix G, "the air leakage rate of the building envelope (I_{75Pa}) at a fixed building pressure differential of 0.3 in. H_2O shall be 0.4 cfm/ft^2."[8] This value, typical of modern energy code-compliant construction, was then converted per Appendix G equations to represent a realistic approximation of an average air leakage resulting from the significantly lower range of pressures expected from typical wind events in the climate zone. The energy model was run with the resulting 0.055 CFM/sq. ft. of wall and then varied to explore the effect of a tighter building enclosure and a leakier building.

INFLUENCE COEFFICIENTS

Influence coefficients are calculated to compare the relative importance of the model assumptions. In this case, influence coefficients were calculated to compare different enclosure performance factors and their effect on the overall energy usage of the building. Influence coefficients are generated for a given baseline data point and are calculated as the partial derivative of the change in the result over the partial derivative of the change in the variable.[9] In effect, this calculation yields the slope of the curve representing the relationship between the variable and the result, at the point of the baseline datum on the curve. Although the relationship between whole-building energy consumption and a given variable is often nonlinear, when comparing data points that are close to the baseline data point on the curve, the slope can be approximated as linear. Because the goal of this paper is to explore the relative importance of different input parameters with dissimilar units, it is necessary to normalize these influence coefficients by dividing the parameter by its base case value and dividing the result by its base case value.[9] This equation can be simplified to the following:

$$\frac{\partial(R^*)}{\partial(P^*)} \approx \frac{\Delta R^*}{\Delta P^*} = \frac{R^*_{bc} - R^*_{\Delta}}{P^*_{bc} - P^*_{\Delta}} = \frac{(R_{bc} - R_{\Delta}) * P_{bc}}{(P_{bc} - P_{\Delta}) * R_{bc}}$$

where:
    P = parameter,
    R = result,
    $*$ = normalized,
    bc = base case, and
    Δ = value for perturbed case.

Once calculated, a normalized influence coefficient for a particular baseline can be used to estimate the percentage error in the result by multiplying by the estimated percentage error in a given assumption. As an example, if a normalized influence coefficient is found to be 0.50 and the estimated error in the input is 5%, then the estimated error in the result can be approximated as $(0.5)*(0.05) = .025$, or 2.5%. Each baseline datum is intended to represent the actual in situ building performance, but because the true in-service performance is not known, assumptions must be made. For this study, the control baseline condition was varied to explore multiple possible "true" performance levels to determine whether the sensitivity of the model is dependent on the magnitude of the performance factor.

The total annual energy usage was considered as the primary energy model result that is used for calculating influence coefficients and to perform comparisons. TRACE 700 parcels the annual energy consumption of the building into several categories, such as lighting, space heating, space cooling, fans, pumps, and miscellaneous loads. Even though varying the building enclosure performance factors primarily affects heating and cooling demand, the true impact cascades into other systems such as the fans and pumps that support these systems. Modifying the envelope resulted in variable run times and efficiencies for the fans and pumps to support the space heating and cooling equipment. Looking at the annual energy consumption differential for the building was a holistic approach to encompass all the affected equipment energies.

Results and Discussion

OPAQUE ASSEMBLY R-VALUE

The goal of varying the thermal resistance of the opaque wall assembly was to determine the sensitivity of the energy model outputs to two different types of wall assemblies frequently encountered in this locality. For this energy model, the opaque assembly makes up about 55% of exterior wall surfaces of the building explored. First, a number of model iterations were generated using the wall assembly approximating an uninsulated mass masonry building and its variants. As one might expect, because thermal resistance was increased the overall energy usage decreased, whereas a reduction in thermal resistance correlated with an increase in the overall energy usage. As the resistance to the transfer of thermal energy into and out of the conditioned building volume is increased at the opaque portions, the result is that the interior conditions are passively tempered, thus decreasing the space conditioning loads that the HVAC systems are required to meet. The HVAC systems are required to operate less frequently, on average, therefore requiring less energy to be expended. Interestingly, this relationship was not found to be linear (**fig. 1A**). In fact, a shift in thermal resistance at the lower end of the spectrum resulted in a greater jump in total energy than an equivalent shift at the higher end of the range. This trend is easily visualized when the influence coefficients are calculated and plotted (**fig. 1B**). As thermal resistance

FIG. 1 Annual energy consumption versus varying mass masonry assembly R-values (A); Influence coefficients of varying mass masonry assembly R-values (B).

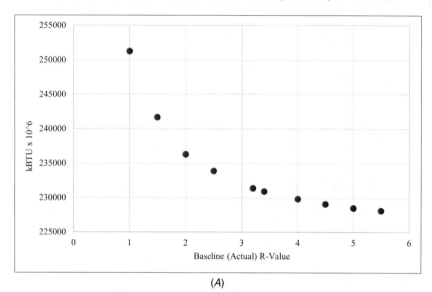

(A)

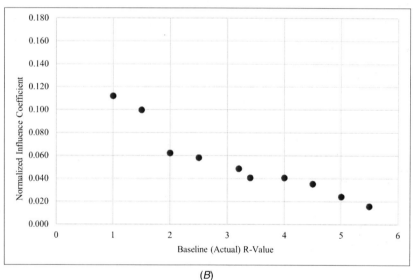

(B)

increases, the influence coefficients drop, indicating a greater model sensitivity to changes when the opaque R-value is lower.

When examining the models generated to represent a more modern metal framed opaque assembly, a similar trend emerged (**fig. 2A**). However, in general the resulting normalized influence coefficients are much lower than those calculated for

FIG. 2 Annual energy consumption versus varying steel-framed assembly R-values (*A*); Influence coefficients of varying steel-framed assembly R-values (*B*).

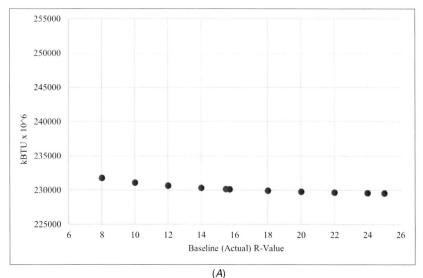

(*A*)

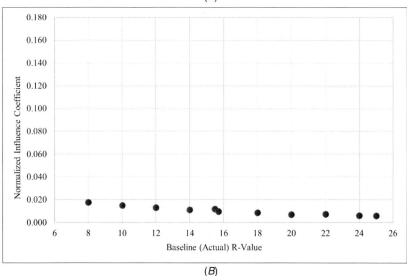

(*B*)

the mass masonry cases (**fig. 2B**). In addition, this set of influence coefficients also illustrates a trend that is much less steep.

Because the relationship of R-value with energy consumption for this case study is nonlinear, it becomes more important to make accurate assumptions regarding the baseline opaque assembly thermal resistivity for lower-performing buildings. In practice, the results of this study demonstrate that field verification of the true in situ thermal resistance of an opaque assembly for the accurate calibration of an energy model becomes increasingly more important when dealing with existing assemblies with lower thermal resistance, such as uninsulated mass masonry buildings. Historically, many buildings constructed in New York City in the first half of the 21st century consist of uninsulated mass masonry and fall into this category.

One field investigation strategy for estimating the R-value of an unknown assembly involves exploratory test cuts at representative locations in the exterior wall assemblies. ASTM E3069, *Standard Guide for Evaluation and Rehabilitation of Mass Masonry Walls for Changes to Thermal and Moisture Properties of the Wall*, in particular Sections 5–7,[10] provide a substantial investigation procedure that may be followed to gain an understanding of and document the exterior wall composition. This includes materials and their arrangement and thicknesses, as well as taking note of air spaces, signs of moisture, potential thermal bridging, etc. Although this procedure is geared toward mass masonry assemblies and is intended to assist in evaluating an appropriate approach for retrofitting a wall, this methodology can be adapted for all types of exterior wall assemblies because it allows for an understanding of the intended function of the existing assembly. The additional benefit of this method is that it can help identify any existing problems or conditions that may become problematic when retrofitting. Once the exterior wall assembly materials and layer thicknesses are documented, the overall thermal resistance of the assembly can be estimated numerically by using engineering judgment to apply published values for thermal conductivity for the various materials encountered. The ASHRAE handbook *Fundamentals* offers a useful bank of typical material properties and discusses methods of calculating the overall R-value.[11]

Although this method of R-value approximation affords a vast improvement over guessing at the thermal properties when nothing is known about the exterior wall contents, the user should be aware that in situ performance can vary significantly from published values. Tabulated material properties and manufacturer data are typically based on laboratory tests performed in controlled environments and ideal steady-state conditions with large temperature differentials and neglect the movement of air and moisture through an assembly.[12] These conditions are often not representative of the real world. In practice, the thermal properties of construction materials are not constant and vary with moisture content and even temperature. Many conditions arising in the field can alter the thermal performance of various materials from published values. For example, air movement that is unaccounted for (whether leakage through an assembly or convection within cavities) can limit the effectiveness of thermal resistance, particularly of batt insulation, which relies on the thermal properties of trapped air. Similarly, the presence of moisture within a wall assembly can lead to an increase in the moisture content of an insulation layer, drastically increasing the thermal conductivity of that layer.

Interestingly, just as models exhibit a higher sensitivity to the accuracy of the lower thermal resistances of uninsulated mass masonry, these assemblies are also likely to vary significantly from their assumed or calculated values. When the possible variability in R-value is considered proportional to its total R-value, considerable inaccuracies can result. In part due to the benefit of their increased thermal mass, these types of walls can significantly outperform their assumed R-values. Masonry walls can absorb and store thermal energy over extended periods of time, later releasing the thermal energy (to both sides of the wall) when the temperature gradient has reduced, thus tempering the rate of transfer of thermal energy and increasing the effective thermal resistivity. In one field study, the authors have measured the actual in-situ thermal resistances of mass masonry assemblies to outperform the theoretical value by between 15% and 150%.

For a higher level of accuracy in determining the true in situ R-value of an assembly, ASTM C1155, *Standard Practice for Determining Thermal Resistance of Building Envelope Components from the In-Situ Data*,[13] should be followed. This process is not always convenient and requires a period of several weeks to months to collect useful data. The calculation process also requires a large temperature differential across the assembly, thus limiting its use under normal space conditioning levels such that it typically must be carried out in the peak of summer or, ideally, in the peak of winter. Still, the process of using instrumentation such as heat-flux sensors to measure the true heat flow into and out of an assembly gives great confidence as to the accuracy of the properties determined.

FENESTRATION PERFORMANCE

Fenestration U-Factor

The goal of varying the thermal performance factors of the fenestrations was to determine the sensitivity of the energy model output to two different thermal performance metrics. First, a series of model iterations were run with the fenestration U-factor varying from the control value of 0.36 BTU/h/sq. ft./°F (**fig. 3A**). The resulting model outputs were used to calculate normalized influence coefficients, which are plotted in **figure 3B**. From these data it is apparent that the sensitivity of the model as it relates to the fenestration thermal performance increases as the U-factor increases. Because thermal conductivity (U-factor) is the inverse of thermal resistivity, this trend is similar to that observed with the opaque assembly. That is to say that when thermal resistance is decreased, the sensitivity of the model is increased.

Much the same as with the opaque assemblies explored, this study shows that due to the sensitivity of the energy model to lower-performing fenestrations, it becomes more important to make accurate assumptions for these types of windows. Lower-performing windows tend to be of older construction, typically wood or steel-framed windows with single pane glazing. Higher-performing windows tend to be found on newer construction (or retrofits of older buildings) and include thermally broken aluminum or composite framing with double- or triple-pane insulated glazing.

FIG. 3 Annual energy consumption of varying fenestration U-factors (*A*); Influence coefficients of varying fenestration U-factors (*B*).

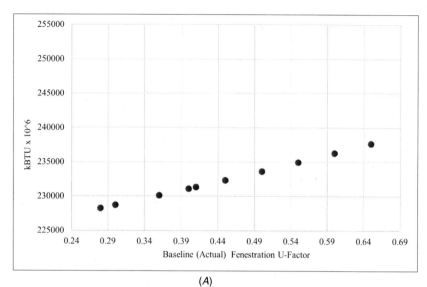

(A)

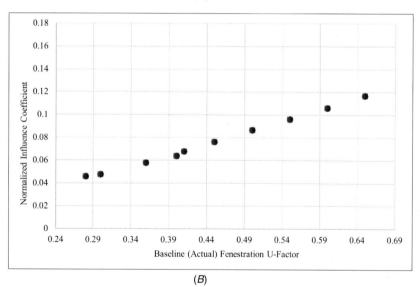

(B)

The first step in determining the performance of a fenestration system is to try to identify the product. Some framing components and most modern glass products will be stamped with product information, sometimes including the make and model. If the product can be identified, it may be possible to locate product data and some thermal properties, although it should be noted that these generic product data are usually specific to a standardized version of the window in brand-new condition. Where the existing product cannot be identified, several standardized methods are available for modeling or measuring the thermal performance of a fenestration. Primarily intended for newly fabricated products that have not yet been installed, these methods seek to establish the typical performance data that will populate a product data sheet. These methods can, however, be adapted to suit existing conditions. NFRC 100, *Procedure for Determining Fenestration Product U-Factors*, establishes a method for two-dimensional thermal simulation to determine the overall U-factor.[14] However, the specific geometries of the framing system must be carefully documented, particularly on extruded elements that can contain hidden stiffeners and screw splines. Limited removal and disassembly of a representative fenestration system may be required to develop an accurate simulation. Other options include NFRC 102, *Procedure for Measuring the Steady-State Thermal Transmittance of Fenestration Systems*; AAMA 1503, *Thermal Transmittance and Condensation Resistance of Windows, Doors and Glazed Wall Sections*; and ASTM C1199, *Standard Test Method for Measuring the Steady-State Thermal Transmittance of Fenestration Systems Using Hot Box Methods*, which offer procedures for the physical measurement of the thermal resistance.[15–17] Due to the challenges associated with adapting the hotbox devices (used to artificially induce a temperature differential in a controlled manner) to field conditions, following one of these standards typically requires the removal of a representative fenestration for testing in a laboratory setting. Other nonstandardized field-investigation methods have been attempted with limited success. In one study, ambient temperature sensors and thermocouples were used to estimate the in situ window performance within 8% of their rated value.[18]

Fenestration SHGC

A number of models were then run to vary the SHGC of the building fenestrations from the control value of 0.4 (**fig. 4A**). The resulting influence coefficients are all larger in magnitude than those associated with the fenestration and opaque U-factors (**fig. 4B**), and there is an overall trend in this data set that would indicate that the sensitivity of the model increases at higher rates of solar radiation transmission.

As discussed previously, the interaction of SHGC with the building's space conditioning loads is complex. Solar radiation acts differently on the various elevations of a building, and some rooms on one side of a building may be well above the set point while rooms on the other side may be below the set point. The location of the thermostat or controls often dictates the actual energy consumption. Although the total energy consumption reliably increases with an increased SHGC, it is likely that this complexity contributes to the observed variability in influence coefficients. The effect of solar radiation also depends greatly on the building elevation and effects of shading from neighboring structures. This study did not take into account shading from adjacent structures due to the lack of information available about the surrounding properties. However, given that this building is located in Midtown

FIG. 4 Annual energy consumption of varying fenestration SHGCs (A); Influence coefficients of varying fenestration SHGCs (B).

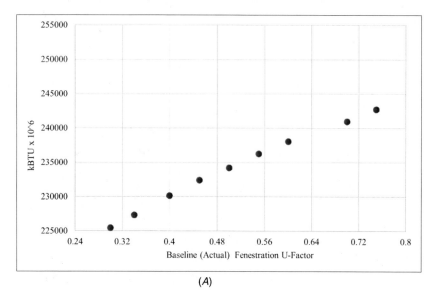

(A)

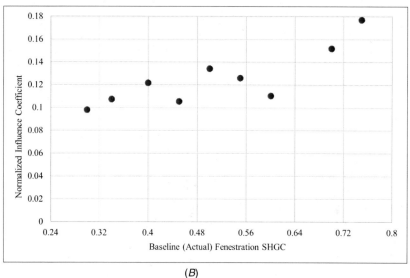

(B)

Manhattan, shading will have a consequential effect on the energy usage, as well as the sensitivity of this metric. Consider that if the first several floors of windows on the north elevation are perpetually in full shade, this is similar to a direct reduction in the fenestration area subject to solar radiation. It is likely that the effect of the solar heat gain demonstrated by this study is an overestimation. However, even when taking into account shading, this factor will be important to overall accuracy and should be explored in as much detail as possible. Just as seen with the opaque assemblies explored and the effect of a window's thermal transmittance, this case shows that the energy model is generally more sensitive to accurate assumptions regarding lower-performing fenestrations, such as single-pane uncoated glass.

One available method to calculate the SHGC of a window is a simulation in accordance with NFRC 200, *Procedure for Determining Fenestration Product Solar Heat Gain Coefficient and Visible Transmittance at Normal Incidence*.[19] As with the simulation procedure described previously, the window system will need to be well documented. For increased accuracy, an investigation could consider performing a site study to measure solar incidence and shading effects at various surfaces of the building over the course of a year to update the model at a more granular level, especially in the absence of data on the neighboring structures.

BUILDING AIR LEAKAGE

The goal of varying the whole building air leakage was to determine the sensitivity of the energy model output to variability in air leakage. As previously reported, a modern code-compliant building would exhibit an air leakage of 0.4 cfm/sq. ft. when tested at 1.57 psf. This air leakage is adjusted to 0.055 cfm/sq. ft. to approximate the air leakage at pressures experienced by the building on a daily basis. A number of iterations of the model were run to explore a series of air leakage values within this order of magnitude. The model results were predictable in that a higher quantity of air leakage resulted in an increase in total energy consumption (**fig. 5A**). When exterior air finds pathways to bypass the environmental control layers, it directly introduces thermal loads into the building spaces that require the HVAC systems to run more to reach interior set points. The calculated influence coefficients (**fig. 5B**) demonstrate that there is a greater sensitivity in the energy model to higher levels of air leakage.

Practically speaking, the results indicate that the overall importance of accuracy in modeling air leakage is greater for looser construction. Air leakage is an important factor because it has the ability to transfer large amounts of moisture bearing a consequential latent heat capacity. Pallin et al. suggested that air leakage, which can be governed by qualitative regulations such as requiring a continuous air barrier, is often overshadowed by other performance factors that are required to meet specific prescriptive quantitative limitations.[20] Yet several standards exist for the field measurement of air leakage, and these tests are becoming more commonplace for the quality control and verification of new building construction. These field testing standards are effective in demonstrating the relative performance of buildings at standardized and artificially induced pressure differentials. However, air leakage is a very complicated phenomenon, and there are significant challenges in translating these data to a realistic prediction of the actual air leakage that results at naturally occurring pressures.

Methods for field determination of air leakage range from tests for specific components (ASTM E783, *Standard Test Method for Field Measurement of Air*

FIG. 5 Annual energy consumption of varying air leakage rates (A); Influence coefficients of varying air leakage rates (B).

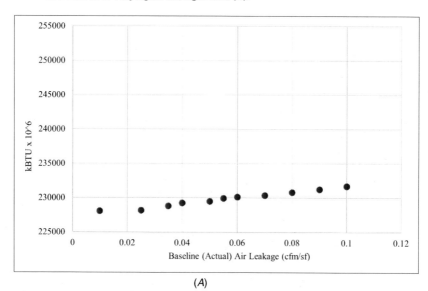

(A)

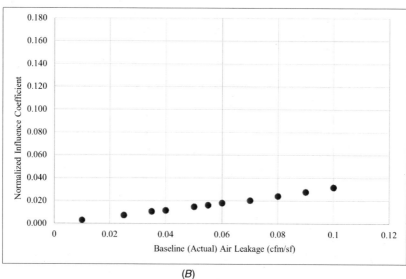

(B)

Leakage through Installed Exterior Windows and Doors)[21] to pressurization of a volume of building space(s) to determine whole building air leakage rates (ASTM E779, *Standard Test Method for Determining Air Leakage Rate by Fan Pressurization*; ASTM E1827, and the U.S. Army Corps of Engineers' *Air Leakage Test Protocol for Building Envelopes*).[22–24] Many of these standardized field testing methods are performed at fixed pressure differentials, such as 75 Pa (1.57 psf) or 300 Pa (6.24 psf). This standardization allows for direct comparison to codified limits and to verify construction quality, but these pressures are at the higher end of the range of pressures that the building will ever experience. For reference, the static pressures 6.24 and 1.57 psf correlate roughly to wind speeds of 50 and 25 mph, respectively. In reality, climatic data for New York City for the last 12 months shows that the average monthly wind speed ranged from 3 to 7 mph, whereas the maximum gust speed in any month never reached 50 mph.[25] It is also important to use sound engineering judgment when performing these field tests to ensure that the test conditions are representative of in-service behavior. Some tests are intended to exclude air leakage of certain elements to isolate the infiltration/exfiltration associated with other elements (such as the mechanical systems and the enclosure) but may not provide accurate results.

On the modeling software side, some programs are limited in their capabilities for inputting air leakage data. Although many programs allow the input of an air leakage schedule that could be developed on the basis of historical wind velocity data and coupled with forced air system operation schedules to estimate actual assembly pressure differentials on an hourly basis, in practice it is difficult to translate historical climatic data into realistic leakage values. Accordingly, some practitioners account for air leakage as a fixed (constant) input. Furthermore, programs such as the one used for this study apply this leakage only during periods in which the HVAC systems are not operating. It can be inferred that the software considers only air infiltration and that when forced air systems are operating the building is positively pressurized, precluding air infiltration. This assumption neglects to consider the effect of air exfiltration, which is maximized when the building is positively pressurized.

It is not realistic to apply a constant value for air leakage. There are currently rules of thumb available for converting air leakage measured at a fixed pressure differential to an estimated air leakage rate to be used for design-level energy modeling. As previously discussed, ASHRAE 90.1 Appendix G provides a formula to estimate this conversion to a constant value.[8] Although this conversion may be useful in approximating the amount of air leakage anticipated for design purposes with new construction, it is unlikely to be accurate enough for the calibration of an energy model of an existing building. In theory, discretizing the effect of air leakage in the energy model should vastly improve the accuracy of the model. However, there are still significant challenges in measuring and estimating air leakage in a manner that is appropriate for energy modeling.

In addition to the variation in wind speeds as a function of time, the effect of wind on a building is highly sensitive to the building location, shape, size, surrounding topography, as well as the particular direction of the wind. ASCE 7, *Minimum Design Loads and Associated Criteria for Buildings and Other Structures*, has developed numerical relationships and formulas for quantifying many of these effects.[26] However, it must be emphasized that these calculations were developed with certain built-in conservatisms because they are intended for structural design purposes. Depending on the incidence of wind, some faces of a building will receive a positive windward load that will vary by height, whereas other faces receive a suction load (**fig. 6**). The simplified load variation depicted in **figure 6** do not account for special behavior at corner zones, overhangs, projections, and parapets. Combined with the fact that the internal pressure in high-rise buildings can vary by elevation due to the stack effect, the specific pressure differential across an assembly can vary greatly in

FIG. 6 Variability of induced wind loads.

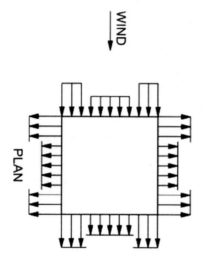

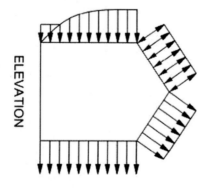

magnitude and direction throughout the exterior surfaces of a building under a single wind load. In addition, the magnitude and direction of the wind change second by second, resulting in an unfathomable level of variation in pressures throughout the exterior during the course of a day. This behavior is vastly different from the conditions measured in the standardized field tests in which air leakage is determined under a single, uniform pressure differential. Last, air leakage does not behave in a one-dimensional manner across a given exterior assembly. The infiltration and exfiltration pathways are dictated by the location of the specific weaknesses in the enclosure layers and may travel through convoluted pathways to pass through an assembly.[27] This effect is exaggerated when assemblies share internal cavities. In other words, air leakage appearing at one location may be the result of a pressure differential across the enclosure at an entirely separate location of the building. There is a need in the industry for additional research to develop methods to improve the accuracy of field-measured air leakage and the conversion of these data to an energy model input. The use of computational fluid dynamics software could be explored to continue to better the understanding of the relationship between wind loading, internal pressurization, and air leakage.

SUMMARY OF RESULTS

Influence coefficients generated for each building enclosure performance factor were normalized for comparison and determination of the relative importance of each. The minimum and maximum normalized influence coefficients for each category are provided in **table 2**. The influence coefficients for several of the performance factors varied widely over the range of inputs. Still, it is possible to draw comparisons in the data, which suggests that the model for this building is most sensitive to the accuracy of the fenestration SHGC. The influence coefficients at

TABLE 2 Summary of normalized influence coefficients

		Normalized Influence Coefficient ($\times 10^{-3}$)	Estimated Percentage Error in Result		
			With 5% Error in Parameter	With 25% Error in Parameter	With 100% Error in Parameter
Opaque R-value—historic uninsulated masonry	Minimum	24.2	0.12%	0.61%	2.42%
	Maximum	139.9	0.70%	3.50%	13.99%
Opaque R-Value—modern metal-framed	Minimum	6.0	0.03%	0.15%	0.60%
	Maximum	22.4	0.11%	0.56%	2.24%
Approximate fenestration U-factor range	Minimum	36.1	0.18%	0.90%	3.61%
	Maximum	105.9	0.53%	2.65%	10.59%
Approximate fenestration SHGC range	Minimum	68.6	0.34%	1.72%	6.86%
	Maximum	151.9	0.76%	3.80%	15.19%
Approximate total air leakage range	Minimum	2.6	0.01%	0.06%	0.26%
	Maximum	31.8	0.16%	0.80%	3.18%

both ends of the spectrum considered for this variable are larger than any other calculated influence factor. The next most important performance factors appear to be the fenestration U-factor and the R-value of a low-performing opaque assembly (i.e., uninsulated mass masonry). However, the relationship with influence coefficients is steeper for the low-performing opaque assembly thermal resistivity, indicating that the fenestration U-value can be expected to be more influential over the ranges considered. Last, whole building air leakage and the R-value of a higher-performing opaque assembly tended to be the least influential variables in the ranges explored. These influence coefficients were an order of magnitude lower than those for fenestration SHGC.

These findings indicate the relative model sensitivity to various building enclosure performance factors and can be used to give an indication of which variables should be prioritized for field measurement and verification to accurately calibrate a model. Yet the influence coefficients alone do not give the full picture. The normalized influence coefficients show the relative sensitivity to the variables explored, but the total magnitude in error for the model result is a function of the influence coefficient as well as the percentage error in the input. **Table 2** provides the estimated total model output error for each calculated influence coefficient at three different levels of input error (5%, 25%, and 100%). For all the reasons previously described, air leakage can be a very challenging variable to estimate or quantify. As an example, if the actual air leakage of a building is at the higher end of the range explored and the model input assumption is off by a factor of 25%, the total model error could be over a half percent, which is more significant than the resulting error for a 5% error in the R-value assumption for any of the opaque wall assemblies considered.

Although some of the resulting errors presented in **table 2** may seem at first glance to be manageable, one must consider that many of the performance factors explored have demonstrated greater sensitivities to lower-performing components. It is at this lower end of the range of performance metrics that modeler assumptions can more easily be off by a great percentage, thus compounding the resulting error (i.e., modeling an assembly as R-4 when it actually performs as R-2 versus modeling an R-12 assembly as R-14). The accuracy of the energy model is also amplified by the significance of the model intent. A 1% error for this building correlates to a difference of millions of kBTUs per year, which depending on the specific fuel sources under Local Law 97 could translate into tens or hundreds of thousands of dollars in fines.

Thus, calculated influence coefficients should be considered together with the estimated order of magnitude in error of their assumed inputs. Opaque wall assembly and fenestration thermal resistance can be estimated or measured with a relatively high accuracy for an energy model input. Fenestration solar heat gain and building air leakage can be measured, but the methods for converting these values to an accurate energy model input are challenging and provide an opportunity for significant error. Additional investigation is recommended for continued improvement in quantifying these performance metrics for energy modeling.

Conclusions

This paper explored the sensitivity of building enclosure performance parameters that can be field-verified or studied to produce more accurate energy models. These energy models can then be used with greater confidence to evaluate the costs and benefits of proposed retrofits to meet specific whole building energy usage limitations. If model input assumptions are inaccurate when calibrating a model to match known utility data, some amount of energy expenditure will be misallocated. This means that the use of the model to predict future energy usage with any proposed enclosure retrofits may not be a reliable indicator of actual future energy use. For example, if active air leakage is underestimated and unimproved, the impact of adding supplementary insulation to opaque assemblies will be hindered. Furthermore, significant long-term durability concerns can arise if this air leakage is not identified or considered in the retrofit design. Legislation, such as Local Law 97 of New York City, has imposed specific energy consumption limitations and imposes fines when these levels are not met. Thus, improving the accuracy of a calibrated energy model for predicting future energy use has become critical.

Field-verification methods can be time-consuming and costly, and some may require destructive openings, but a focused field investigation can provide a significant benefit and value to a retrofit project. The evaluation of normalized influence coefficients assists in identifying building performance factors that should be prioritized in such a field investigation to improve model accuracy. The methods prescribed in ASTM E3069[10] are recommended not only for improving model accuracy but also assisting with the development of responsible enclosure retrofit strategies that consider the long-term durability of the structure and assemblies. This standard is intended primarily for mass masonry assemblies but can be adapted for almost any opaque assembly.

A direct comparison of the normalized influence coefficients calculated in this study indicates that the energy model was most sensitive to the accuracy of the fenestration SHGC. The fenestration U-factor and low-performance opaque R-value were slightly less influential. The least influential inputs in the ranges considered were that of a higher-performing opaque wall R-value and the total air leakage. It was also noted that the total error in the model is dependent not only on the sensitivity to a particular input but also the magnitude of error in the input assumption. Therefore, a parameter exhibiting a relatively small influence coefficient can lead to a greater magnitude of error in the model result if the error in the input is great enough.

Normalized influence coefficients for the opaque assembly R-value were found to decrease with increasing thermal resistivity. Accordingly, the verification of in situ R-values is more critical for lower-performing, uninsulated mass masonry assemblies because the data show the model can be five times more sensitive to assemblies in this range. One basic method includes exploratory test cuts to document and measure the composition of the assembly and calculating an estimated thermal resistance on the basis of published material properties. These types of assemblies have been known to exhibit a proportionally large deviation from the theoretical R-value, and measuring the actual R-value of the entire assembly with heat-flux sensors is recommended for greater accuracy.

Fenestration U-value influence coefficients followed a similar trend to that of the opaque assemblies in that the model was more sensitive to lower-performing window systems. Standardized simulation procedures are available for estimating the fenestration thermal performance, although meticulous documentation of the fenestration system geometry is required when product information is not readily available. For increased accuracy, a representative fenestration may be removed to a laboratory setting to measure the thermal resistance of the system using a standardized procedure.

The performance metric with the greatest influence on the model accuracy was found to be fenestration SHGC. Despite some variability in the influence coefficients as a function of the SHGC for the range considered, the model was generally more sensitive to fenestrations with higher SHGCs. However, it is likely that the effect of varying the SHGC for all fenestrations on the building is overestimated by this analysis as a result of the inability to accurately quantify the shading effects of the dense Midtown Manhattan neighborhood. The specific SHGC of a fenestration can be simulated using field measurements of the window system when product data are not available, yet the challenges in accurately representing the significant shading effects in a location such as New York City persist.

This study indicates that model sensitivity to air leakage may not be as significant as other building enclosure performance factors. Yet air leakage is a very complicated phenomenon and is one of the most difficult performance factors to correlate field measurement with naturally occurring infiltration and exfiltration for input to a model. Therefore, the magnitude of error in assumptions regarding inputs for air leakage are anticipated to be much higher. Several standardized test methods exist for the measurement of air leakage at controlled conditions, but there is a need in the industry to better understand the magnitude of error that can actually be expected when using conversion formulas for measurements taken under standardized test methods. Due to the complexity of the phenomenon of air leakage and its relationship to wind speed and direction, it is unlikely that existing conversion formulas are very accurate for the majority of existing buildings.

These influence coefficients also provide an understanding of the sensitivity of building enclosure retrofits to certain upgrades. As an example, the data clearly show that there is a diminishing return on improving the R-value of an opaque assembly or the U-factor of a fenestration. Relatively speaking, a retrofit for this particular building geometry should prioritize improving the solar heat gain performance of the fenestrations and improving the R-value of uninsulated mass masonry assemblies.

The focus of this study was how building enclosure performance factors impact a building's energy use. For consistency, whereas the enclosure performance was varied, the mechanical systems were assumed to remain the same. However, when retrofitting an existing building it is also important to understand how the envelope impacts the heating and cooling loads of the space. With reduced heating and cooling loads within an existing building, there is a potential to downsize the HVAC systems to meet the new loads. For AC units especially, the smaller the unit, the higher its efficiency. Replacing AC units that have reached the end of useful life with smaller AC units could result in further energy savings.

This study explored only one building as a single reference case, and these influence coefficients are likely to vary for every building based on its location and configuration (exterior wall area to conditioned volume, window-to-wall ratio, etc.). When presented with a different existing building that requires a calibrated energy model, a simplified version of this exercise could be performed. After generating the model geometry and initial assumptions, influence coefficients could be developed for the specific building and guide the development of a field investigation to update model assumptions. When performing an investigation to quantify these performance inputs, field measurements obtained typically yield a measurement for a single location on the building at one point in time. These field methods can be time-consuming and costly and may require destructive methods; thus, it is desirable to limit the amount of data points that must be collected only to those that are essential. It is important for field studies to take representative samples and measurements using good engineering judgment to gather a sufficient sampling across different elevations, floor levels, exposures, assemblies, time of day, and so on.

Recommendations for Further Studies

As done in this study, a similar sensitivity analysis could be performed for any number of additional building types, including buildings with different aspect ratios, climates, ratio of exterior enclosure area to conditioned volume, window-to-wall ratios, and more.

There are dozens of energy modeling programs deployed in the industry today. Although TRACE 700 was used for this study, it could be valuable to test the same existing building and building performance factors using a different modeling program. TRACE 700 is limited in its ability to graphically depict the building's geometry in three dimensions and requires lengthy run times for a building just under 2 million sq. ft. Trane launched a new modeling program in October 2020 called Trane 3D Plus that is an EnergyPlus^{TM}-based modeling program. EnergyPlus code was developed by the U.S. Department of Energy and is also considered robust for building simulation applications. It would be worthwhile to rerun the conditions of this paper using Trane 3D Plus or another EnergyPlus-based program.

Additional performance factors that could be evaluated include thermal bridging and modeling techniques for fenestration perimeter conditions. As typical energy modeling programs accept only a single R-value for a given assembly type in the library, it is important to identify and quantify thermal bridging. Regularly spaced thermal bridges such as framing can easily be accounted for by using the parallel path method[11] to calculate an effective R-value. However, discrete thermal bridges should also be accounted for if they are significant. Further studies should consider performing THERM simulations to refine the effective R-values of assemblies and exploring the sensitivity of an energy model to the accurate quantification of thermal bridging. THERM is a two-dimensional thermal simulation program designed to analyze the heat transfer through complex configurations of building components. Similarly, in the model for this study, an adiabatic condition was assumed between the fenestration and wall assemblies, but Bhandari et al. reported that when including adjacent opaque assemblies in a THERM model for a window wall, the assembly U-factor can increase by 10%.[28] Further studies should consider exploring the effect of modifying this perimeter condition to include a correction factor.

Air leakage accounts for roughly 6% of total energy use in commercial buildings within the United States.[29] As dissected throughout this study, software capabilities to model this parameter are often limited to one or two inputs with a given schedule. In reality, air leakage is a complicated event often with variability unto itself.[30] Oak Ridge National Laboratory in collaboration with several other organizations is working to develop a stand-alone program to calculate the air leakage of a building on a more granular level, considering variables such as climate and stack effect, and with the capability to feed this information directly into an energy model.[29] Although this method develops a framework for greater capability to account for air leakage in the model, it still relies upon theoretical approximation. Further studies should look to bridge the gap between this energy modeling method and field measurement of air leakage to maximize the accuracy of this variable.

References

1. The City of New York, Mayor's Office of Long-Term Planning and Sustainability, "One City Built to Last," 2014, http://web.archive.org/web/20210221211111/http://www.nyc.gov/html/builttolast/assets/downloads/pdf/OneCity.pdf
2. New York City Council, "Local Laws of the City of New York for the Year 2019: No. 97," 2019, http://web.archive.org/web/20220104211943/https://www1.nyc.gov/assets/buildings/local_laws/ll97of2019.pdf
3. Urban Green Council, "NYC Building Emissions Law Summary: Local Law 97," 2020, https://www.urbangreencouncil.org/sites/default/files/urban_green_building_emissions_law_summary_2020.02.19.pdf
4. G. Chaudhary, J. New, J. Sanyal, P. Im, Z. O'Neill, and V. Garg, "Evaluation of 'Autotune' Calibration against Manual Calibration of Building Energy Models," *Applied Energy* 182 (2016): 115–134, https://doi.org/10.1016/j.apenergy.2016.08.073
5. K. Soratana and J. Marriott, "Increasing Innovation in Home Energy Efficiency: Monte Carlo Simulation of Potential Improvements," *Energy and Buildings* 42, no. 6 (2010): 828–833, https://doi.org/10.1016/j.enbuild.2009.12.003

6. H. Brohus, P. Heiselberg, A. Hesselholt, and H. Rasmussen, "Application of Partial Safety Factors in Building Energy Performance Assessment" (paper presentation, Building Simulation 2009, Glasgow, Scotland, July 27–30, 2009).
7. Trane, *TRACE® 700 Building Energy and Economic Analysis User's Manual, Version 6.2* (La Crosse, WI: Author, 2010).
8. *Energy Standard for Buildings Except Low-Rise Residential Buildings (I-P Edition)*, ASHRAE Standard 90.1 (2016) (Atlanta, GA: ASHRAE, 2016).
9. J. D. Spitler, D. E. Fisher, and D. C Zietlow, "A Primer on the Use of Influence Coefficients in Building Simulation" (paper presentation, Building Simulation 1989, Vancouver, Canada, June 23–24, 1989).
10. *Standard Guide for Evaluation and Rehabilitation of Mass Masonry Walls for Changes to Thermal and Moisture Properties of the Wall*, ASTM E3069-19a (West Conshohocken, PA: ASTM International, approved October 1, 2019), http://doi.org/10.1520/E3069-19A
11. ASHRAE, *Fundamentals* (Atlanta, GA: Author, 2017).
12. *Standard Test Method for Thermal Performance of Building Materials and Envelope Assemblies by Means of a Hot Box Apparatus*, ASTM C1363-19 (West Conshohocken, PA: ASTM International, approved September 1, 2019), http://doi.org/10.1520/C1363-19
13. *Standard Practice for Determining Thermal Resistance of Building Envelope Components from the In-Situ Data*, ASTM C1155-95(2013) (West Conshohocken, PA: ASTM International, approved November 1, 2013), http://doi.org/10.1520/C1155-95R13
14. *Procedure for Determining Fenestration Product U-Factors*, ANSI/NFRC 100-2020 (Greenbelt, MD: National Fenestration Rating Council, 2020).
15. *Procedure for Measuring the Steady-State Thermal Transmittance of Fenestration Systems*, NFRC 102-2020 (Greenbelt, MD: National Fenestration Rating Council, 2020).
16. *Thermal Transmittance and Condensation Resistance of Windows, Doors, and Glazed Wall Sections*, AAMA 1503-09 (Schaumburg, IL: AAMA, 2009).
17. *Standard Test Method for Measuring the Steady-State Thermal Transmittance of Fenestration Systems Using Hot Box Methods*, ASTM C1199-14 (West Conshohocken, PA: ASTM International, approved February 1, 2014), http://doi.org/10.1520/C1199-14
18. K. Varshnay, J. E. Rosa, and I. Shapiro, "Method to Diagnose Window Failures and Measure U-Factors on Site," *International Journal of Green Energy* 9 (2012): 1–17.
19. *Procedure for Determining Fenestration Product Solar Heat Gain Coefficient and Visible Transmittance at Normal Incidence*, ANSI/NFRC 200-2020 (Greenbelt, MD: National Fenestration Rating Council, 2020).
20. S. Pallin, M. Bhandari, and M. Lapsa, "Quantifying Hidden Value with the Building Enclosure Performance Metric" (paper presentation, Building Envelope Technology Symposium, Nashville, TN, November 16–17, 2018).
21. *Standard Test Method for Field Measurement of Air Leakage through Installed Exterior Windows and Doors*, ASTM E783-2(2018) (West Conshohocken, PA: ASTM International, approved October 1, 2019), http://doi.org/10.1520/E0783-02R18
22. *Standard Test Method for Determining Air Leakage Rate by Fan Pressurization*, ASTM E779-19 (West Conshohocken, PA: ASTM International, approved January 1, 2019), http://doi.org/10.1520/E0779-19
23. *Standard Test Method for Determining Airtightness of Buildings Using an Orifice Blower Door*, ASTM E1827-11(2017) (West Conshohocken, PA: ASTM International, approved September 1, 2017), http://doi.org/10.1520/E1827-11R17
24. U.S. Army Corps of Engineers, *Air Leakage Test Protocol for Building Envelopes*, Version 3 (Champaign, IL: U.S. Army Engineer Research and Development Center, 2012).

25. National Oceanic Atmospheric Administration, "NY City Central Park Weather Data, NY US," 2021, https://web.archive.org/web/20210315023159if_/https://www.ncei.noaa.gov/access/search/data-search/global-hourly?datasetabbv=DS3505&countryabbv=US&georegionabbv=&resolution=40&dataTypes=WND&bbox=40.965,-74.257,40.465,-73.757&startDate=2020-01-01T00:00:00&endDate=2020-12-31T23:59:59
26. *Minimum Design Loads for Buildings and Other Structures*, ASCE/SEI 7-10 (Reston, VA: ASCE, 2013), http://doi.org/10.1061/9780784412916
27. J. W. Lstiburek, K. Pressnail, and J. Timusk, "Transient Interaction of Buildings with HVAC Systems—Updating the State of the Art," *Journal of Building Physics* 24 (2000): 111–131.
28. M. Bhandari and R. Srinivasan, "Window-Wall Interface Correction Factors: Thermal Modeling of Integrated Fenestration and Opaque Envelope Systems for Improved Prediction of Energy Use" (paper presentation, Fifth National Conference of IBPSA-USA, Madison, WI, August 1–3, 2012).
29. M. Bhandari, D. Hun, S. Shrestha, S. Pallin, and M. Lapsa, "A Simplified Methodology to Estimate Energy Savings in Commercial Buildings from Improvements in Airtightness," *Energies* 11 (2018): 3322, https://doi.org/10.3390/en11123322
30. D. Costola, B. Blocken, M. Ohba, and J. L. M. Hensen, "Uncertainty in Airflow Rate Calculations Due to the Use of Surface-Averaged Pressure Coefficients," *Energy and Buildings* 42 (2010): 881–888, https://doi.org/10.1016/j.enbuild.2009.12.010

STP 1635, 2022 / available online at www.astm.org / doi: 10.1520/STP163520210033

Andrew W. Wagner[1] and Jodi M. Knorowski[1]

Coordination of ASTM Standards for Thermal Insulation Materials Installed in Exterior Cavity Wall Applications

Citation

A. W. Wagner and J. M. Knorowski, "Coordination of ASTM Standards for Thermal Insulation Materials Installed in Exterior Cavity Wall Applications," in *Building Science and the Physics of Building Enclosure Performance: 2nd Volume*, ed. D. J. Lemieux and J. Keegan (West Conshohocken, PA: ASTM International, 2022), 29–52. http://doi.org/10.1520/STP163520210033[2]

ABSTRACT

ASTM standards have been and continue to be developed to provide criteria for material classification and establish best practices for building envelope components. They provide the test methods needed to evaluate the behaviors and properties of different materials such that these materials can be compared and contrasted. However, the wide variety of materials and material applications for the building envelope can make the application of ASTM standards for specific project conditions challenging for designers and industry professionals to properly assess. The standardization process creates consistency in the methods used to identify and define material properties; however, the test methods used to evaluate a specific type of material may differ among materials that could be used for similar applications. As a result, this limits the ability of the design professional to accurately compare different materials that would be installed in the same application in order to select the most appropriate overall assembly to achieve the goals for building performance. When evaluating the thermal performance of the building envelope, there are several types of thermal insulation products that are commonly used in exterior cavity wall applications,

---

Manuscript received March 17, 2021; accepted for publication August 16, 2021.
[1] WDP & Associates Consulting Engineers, Inc., 335 Greenbrier Dr., Charlottesville, VA 22901, USA A. W. W. http://orcid.org/0000-0002-4649-5024, J. M. K. http://orcid.org/0000-0002-9170-6839
[2] ASTM Second Symposium on *Building Science and the Physics of Building Enclosure Performance* on April 24–25, 2022, and June 12, 2022 in Seattle, WA, USA.

Copyright © 2022 by ASTM International, 100 Barr Harbor Drive, PO Box C700, West Conshohocken, PA 19428-2959.

ASTM International is not responsible, as a body, for the statements and opinions expressed in this paper. ASTM International does not endorse any products represented in this paper.

yet the test methods contained within the standard specifications for each insulating material are not entirely coordinated. Although the methods used to determine thermal resistance are generally consistent between different material types, impacts of moisture and dimensional stability are often evaluated differently, which can be misleading when comparing materials for the same application. This paper presents the challenges that are faced by the design professional when evaluating different materials on the basis of the current guidance provided in ASTM standards and recommendations for methods to coordinate requirements for materials installed in similar applications.

Keywords

thermal insulation, specifications, test methods, building envelope performance

Introduction

Within the building envelope, materials are specified to generally serve a primary purpose. However, the properties of that material can allow it to function in multiple ways. The insulation layer of the building envelope is typically intended to provide thermal resistance across a wall assembly. Depending on the type of insulation specified, other physical properties also become a factor in the overall design and performance of the wall assembly, such as vapor permeance and moisture resistance. When comparing different types of materials that can be specified for the same primary purpose, it is important to understand the other material properties to ensure the designed assembly functions as intended.

For each type of building material, there is generally an ASTM standard specification that outlines the physical properties and applicable test methods that standardize the performance requirements such that each type of material can be compared equally to other similar types of materials. There are many different standard test methods that can be used to evaluate the physical properties of a material. Some of these test methods aim to evaluate similar properties of the material; however, the differences in the test setup, duration, or approach can create vastly different results. Within the industry, the reporting of the results of these tests is concise so that they can be presented on a product data sheet in a manner that is easy to review by the end user. This simplicity can provide an incomplete summary of the material properties, especially when comparing different types of materials that may be tested in different ways to evaluate similar material properties. Furthermore, some material specifications may not require certain tests to be performed, leaving a wider gap in being able to accurately compare the performance of different materials installed in the same application.

Various insulating materials are available within the industry that can be used to create the building thermal envelope. This paper focuses on the comparisons of standard specifications between four common insulation types that are often installed in an exterior cavity wall assembly: extruded polystyrene (XPS) insulation, mineral fiber insulation, polyisocyanurate insulation, and sprayed polyurethane foam (SPF) insulation. Standard specifications for some of these thermal insulation materials were first developed in the 1980s. Over time, these standards have been revised, and new material specifications have been developed as technology has led to the creation of new materials. Although the test methods for determining the thermal performance of insulation are similar, test methods for determining other material properties can vary between types of insulation. This paper outlines some of the discrepancies with the published thermal insulation standard specifications and discusses the potential impacts this can have on the design and performance of the end-use condition within a wall assembly.

Standard Specifications

XPS INSULATION

XPS insulation is specified in ASTM C578, *Standard Specification for Rigid, Cellular Polystyrene Thermal Insulation*.[1] Performance requirements outlined in this standard include compressive strength, thermal resistance, flexural strength, water vapor permeance, water absorption, and dimensional stability in order to classify the type of XPS insulation. It is noted within the specification that XPS insulation is combustible, and although the oxygen index is determined for each type of insulation, this is not intended to be a measure of fire risk but more generally relates to products manufactured with and without flame retardants.

Generally, test specimens that are evaluated in each of the specified test methods are 1 in. thick. Sampling is required to be performed in accordance with ASTM C390, *Standard Practice for Sampling and Acceptance of Thermal Insulation Lots*.[2] Criteria for selecting products for testing during the manufacturing process to determine classifications for the products and provide quality control are outlined in this standard. The requirements for conditioning requirements are referenced in the specification to be in accordance with ASTM C870, *Standard Practice for Conditioning of Thermal Insulating Materials*,[3] for each test method to ensure any impacts from moisture exposure, temperature exposure, or aging of the material would be consistent at the start of each test. For thermal resistance, the classification of the insulation is based on a mean temperature of 75 ± 2°F. Certain types of XPS insulation are required to report the long-term thermal resistance (LTTR), but this property is not required for classification of the material. For foam plastic insulation, the thermal performance generally relies on blowing agents that are captured within the foam during the manufacturing process. Over time, these blowing agents will dissipate, which will decrease the thermal performance of the insulation. The LTTR provides the thermal resistance of the insulation when it has been aged for 5 years, which corresponds to the thermal resistance over a 15-year service life. Within the Appendix, which includes nonmandatory information, considerations are provided for the end use of the insulation. As it relates to thermal resistance, values are provided for each type of insulation at additional mean temperatures of 25, 40, and 110°F. It is also noted that extended water absorption, as well as exposure to freeze/thaw cycles, could have a negative impact on the thermal performance of the material. Within the Appendix, the water vapor transmission is discussed in relation to adjacent vapor-retarding components, noting that the vapor permeance of the XPS should be considered when developing specifications for the vapor retarder.

MINERAL FIBER INSULATION

Mineral fiber insulation is specified in ASTM C612, *Standard Specification for Mineral Fiber Block and Board Thermal Insulation*.[4] The performance requirements outlined in this standard to classify the type of mineral fiber insulation include the compressive resistance, thermal conductivity, water vapor sorption, surface burning characteristics, and linear shrinkage. In addition to the classification of the various types of insulation, test methods are also included that require the mineral fiber to be tested for odor emission, corrosiveness to steel, and fungi resistance. Various facing materials that can be applied to the mineral fiber are provided within the specification, but specific guidance is not provided on the selection of such facings, other than referencing another standard for low vapor permeance facings (ASTM C1136, *Standard Specification for Flexible, Low Permeance Vapor Retarders for Thermal Insulation*[5]).

Sampling of the insulation is required to be in accordance with ASTM C390. A single thickness of the mineral fiber insulation test specimen is not recommended for each test outlined in the specification. There are also no specific conditioning requirements that are outlined in the specification. Within the material specification, the test specimen sizes and conditioning requirements defer to the requirements outlined in each of the specified test methods.

POLYISOCYANURATE INSULATION

Polyisocyanurate insulation is specified in ASTM C1289, *Standard Specification for Faced Rigid Cellular Polyisocyanurate Thermal Insulation Board*.[6] The performance requirements outlined within the specification to classify the type of polyisocyanurate insulation include the compressive strength, tensile strength, flexural strength, thermal resistance, water vapor permeance, water absorption, and dimensional stability. Various facers, which are adhered materials such as felt, paper, or glass, that can be incorporated on one or both sides of the board are outlined within the specification, with general performance requirements for each type of facer. The LTTR is also included, with guidance on preparing the test specimens; however, there are several acceptable ways to prepare the test specimens that could cause discrepancies in the test results between different products. Polyisocyanurate insulation is noted to be combustible, but methods for determining surface burning characteristics are provided, if required.

Samples for testing are required to be in accordance with ASTM C390 unless otherwise noted. General conditioning requirements are provided unless specific conditioning parameters are outlined within the individual test method. For thermal resistance, the test specimens are tested at a mean temperature of 75°F. The thermal resistance can also be determined at mean temperatures of 40°F and 110°F at the buyer's request, but these results are not required to "establish compliance with this specification." It is also noted that the thermal properties of the insulation may vary from the reported values on the basis of the end-use condition and how much the installed conditions vary from the test conditions under which the test specimen was tested when determining the thermal properties of the insulation.

SPF INSULATION

SPF insulation is specified in ASTM C1029, *Standard Specification for Spray-Applied Rigid Cellular Polyurethane Thermal Insulation.*[7] The performance requirements that are outlined in the specification in order to classify the type of SPF insulation include compressive strength, tensile strength, thermal resistance, water absorption, water vapor permeability, dimensional stability, and closed-cell content. While SPF insulation is combustible, the surface burning characteristics of the material are required to be reported, although they are not required for classification. SPF is produced by a chemical reaction of mixing several compounds using commercial spray equipment at the time of installation. Because of this, the standard specification includes an Appendix in which the material properties of the product are noted to potentially vary on the basis of conditions at the time of installation.

Detailed information about sampling and preparing test specimens of SPF to perform each of the required tests are included in the specification. Sampling requirements are not based on a specific ASTM standard but rather are agreed upon by the manufacturer and the purchaser. Because SPF is essentially manufactured on a project site, test panels must be created that are consistent with the methods that will be used to install the SPF on the building. Conditions required for curing the SPF following installation are provided, but specific conditioning requirements prior to testing for each physical property are not explicitly outlined. Another unique feature of SPF compared with other types of insulation is the skin coat that is created at the surface of the insulation after installation. Depending on the number of layers that are required to build up the insulation to the specified thickness, knit lines may also be integral to the material that could impact the material properties. Each test method has various requirements for whether or not these skin coats and knit lines should be included as part of the test specimen.

KEY COMPARISONS

When reviewing the material specifications for each type of insulation, the physical properties that are tested for classification of the materials can be categorized as follows: thermal transmission, moisture resistance, water vapor permeance, strength, and dimensional stability. Different test methods are generally specified for each type of insulation for determining the performance of the insulation within each of these categories. There are also requirements within some specifications that are unique to that type of insulation. The physical properties and general approach to testing each property for each type of insulation are summarized in **table 1**.

TABLE 1 Comparison of physical properties for each type of insulation

Physical Property	XPS (ASTM C578)	Mineral Fiber (ASTM C612)	Polyisocyanurate (ASTM C1289)	SPF (ASTM C1029)
Thermal transmission	Resistance at 75°F mean temperature; LTTR for some types of XPS (ASTM C177, C518, C1114, C1363)	Conductivity at range of mean temperatures (ASTM C177, C518, C1114)	Resistance at 75°F mean temperature (ASTM C177, C518, C1114, C1363); LTTR (ASTM C1303)	Resistance at 75°F mean temperature (ASTM C177, C518, C1363)
Moisture resistance	Water absorption (24 h) (ASTM C272)	Water vapor sorption (ASTM C1104)	Water absorption (2 h) (ASTM C209)	Water absorption (96 h) (ASTM D2842)
Water vapor permeance	Desiccant method (ASTM E96)	N/A	Desiccant method (ASTM E96)	Desiccant Method (ASTM E96)
Strength	Compressive resistance (ASTM C165), flexural srength (ASTM C203)	Compressive resistance (ASTM C165)	Compressive strength (ASTM D1621), flexural strength (ASTM C203), tensile strength (ASTM C209)	Compressive strength (ASTM C165), tensile strength (ASTM D1623)
Dimensional stability	Percent linear change (ASTM D2126)	Percent linear shrinkage in length (ASTM C356)	Percent linear change in thickness and length/width (ASTM D2126)	Percent linear change in thickness, length, or width (ASTM D2126)
Other	Oxygen index (ASTM D2863)	Surface burning characteristics (ASTM E84), odor emission (ASTM C1304), corrosiveness to steel (ASTM C665), fungi resistance (ASTM C1338)	Surface burning characteristics (if required) (ASTM E84)	Surface burning Characteristics (ASTM E84), closed-cell content (ASTM D6226)

Standard Test Methods for Material Properties

THERMAL TRANSMISSION

Although several test methods that are acceptable for testing thermal transmission of a material are outlined in each standard specification, many manufacturers default to using ASTM C518, *Standard Test Method for Steady-State Thermal Transmission Properties by Means of the Heat Flow Meter Apparatus.*[8] This test method determines the thermal conductance by sandwiching the test specimen between a hot plate and a cold plate and enclosing the edges to minimize heat loss. Temperature sensors and heat flux sensors are mounted to the surface of the specimen, and when thermal equilibrium is achieved, the temperature difference across the specimen along with the measured rate of heat flow through the specimen is used to determine the thermal conductance. A summary of the required thermal performance for each type of insulation, along with specified thickness, facer requirements, and conditioning requirements, is provided in **table 2**.

Guidance is provided in ASTM C1058, *Standard Practice for Selecting Temperatures for Evaluating and Reporting Thermal Properties of Thermal Insulation*,[9] on how to select temperatures that are required to evaluate the thermal properties of insulation. It is assumed that in moderate climates insulation installed in the building envelope will be exposed to temperatures that range from 30 to 120°F; therefore, it is recommended to test the insulation at mean temperatures of 40, 75, and 110°F. The temperature difference between the hot and cold side of the test apparatus is also provided in this standard. When applying ASTM C1058 to ASTM C518, a small temperature difference of 50 ± 10°F or a large temperature difference of 80 ± 20°F is indicated. For each insulation material, the required mean temperature between the hot plate and cold plate is 75°F, with testing at other mean temperatures upon request. When looking at product data sheets that are readily available in the industry, testing at other mean temperatures is rarely reported. Furthermore, only the polyisocyanurate and SPF specifications indicated a minimum temperature difference of 40°F. Although it is likely that other types of insulation are tested at the same manner, the lack of standardization in reporting results could result in comparisons between materials that are not consistent. In colder climates, the need for understanding the performance of insulation at lower mean temperatures is critical for overall building performance. The temperature differentials between interior and exterior spaces are also much greater in the winter months, thus requiring the building envelope to perform as designed to reduce the loads on the mechanical systems that use energy. If the building design relies on thermal performance reported for a mean temperature of 75°F yet the building is expected to perform at

TABLE 2 Thermal performance specifications for insulation types

	XPS (Type IV)	Mineral Fiber (Type IVB)	Polyisocyanurate (Type I)	SPF (Type II)
Specimen thickness	1 in.	Maximum 2 in.	1 in., 1.5 in., 2 in.	1-in. core sample
Facing	N/A	Unfaced	Not indicated	Remove skins on each side of sample
Conditioning requirements	In accordance with test procedure; per ASTM C518, 71.6°F at 50% RH	None provided; per ASTM C518, 71.6°F at 50% RH	73°F and 50% RH for 180 days or 140°F dry heat for 90 days	73°F and 50% RH for 180 days or 140°F dry heat for 90 days
Thermal resistance (F-ft.^{2}-h/BTU)	R-5 at 1″ (75°F)	R-4.3 at 1″ (25°F) R-4.2 at 1″ (75°F) R-4.0 at 1″ (100°F)	R-6.6 at 1″ (40°F) R-6.0 at 1″ (75°F) R-5.4 at 1″ (110°F)	R-6.2 at 1″ (75°F)

mean temperatures closer to 40°F, these data should be readily available such that designers can specify materials that will meet the performance goals of the building.

The various available test methods that can be used to determine thermal transmission through the insulation should also be considered as it relates to the end use of the product in an exterior wall cavity. Generally, the insulation will be installed tight against the exterior sheathing, and an air space will be present behind the cladding material. When testing thermal transmission in accordance with ASTM C518, the insulation is in direct contact with the hot and cold plates, which does not account for any convective air movement that may occur during the end-use installation in an exterior wall cavity. ASTM C1363, *Standard Test Method for Thermal Performance of Building Materials and Envelope Assemblies by Means of a Hot Box Apparatus*,[10] which is another test method for determining thermal performance, utilizes a hot box apparatus to evaluate the thermal transmission through building materials. The same theory is applied with measuring the temperature differential and rate of heat flow through the insulation, except the test specimen is mounted within a chamber that has a cold side and a warm side, allowing for air circulation at each side of the specimen. This test method may provide a more accurate representation of the thermal performance of the insulation materials based on their installed configuration, although it is generally more time-consuming and costly to perform. It was noted that the specification for mineral fiber insulation did not list ASTM C1363 as an acceptable method for the determination of thermal performance. Because of the fibrous nature of this material, the incorporation of air movement around and through the specimen may not accurately represent the actual thermal performance of the material. To better understand the performance of the insulation material during the end-use condition, requiring the material to be tested in accordance with ASTM C1363 could provide better data for consideration by design professionals; however, it would need to be tested as part of an assembly representative of the end use of the product.

In modern wall assemblies, insulation is often installed outboard of the air and water barrier where it is exposed to environmental conditions that are more representative of exterior conditions than interior conditions, which includes bulk water and a variety of relative humidity levels. The conditioning of the test specimens prior to testing for thermal performance requires exposure to 50% relative humidity at a temperature ranging from 71 to 73°F. When installed, the insulation will be exposed to a much greater range of conditions. As buildings become higher-performing, designers must be able to understand the performance of the building thermal envelope under a variety of conditions. The impacts of moisture on the thermal performance of various types of insulation should be incorporated into standard specifications so that this information is readily available to design professionals.

In addition to evaluating the moisture-dependent thermal resistance, consideration should also be given to the long-term impacts of environmental exposure on thermal performance. A test method to evaluate the effect of environmental cycling on thermal performance of insulation is outlined in ASTM C1512, *Standard Test Method for Characterizing the Effect of Exposure to Environmental Cycling on Thermal Performance of Insulation Products*.[11] For this test, a test specimen is mounted between a warm chamber (75°F and 90% relative humidity) and a cycling chamber (cycles of 5°F and 60°F) for a period of 20 days, and thermal performance is calculated before and after this exposure. Although the conditions of exposure may be more extreme than what the insulation would experience in service, this standard could provide the groundwork for evaluating the impact of moisture on thermal insulation and could be adapted to apply to insulation installed in exterior wall assemblies. Currently, this test method is only referenced in the material specification for polystyrene insulation and could provide valuable insight into the service life of all insulation types as well as the impact of moisture freezing and thawing during the cycling process.

MOISTURE RESISTANCE

While the thermal performance of the insulation is the primary purpose of the material in the building thermal envelope, when installed in the exterior wall cavity, the ability to resist moisture is one of the most important performance indicators of the material. If a material begins to deteriorate or lose its insulating value when exposed to moisture, then the primary purpose of the material cannot be achieved. Each of the insulation types studied in this paper specifies a different test method to evaluate moisture resistance, making it difficult to compare the performance of each type of material. Because this is one of the critical performance characteristics of the material, there is a need in the industry to standardize the evaluation of moisture resistance such that performance impacts can be understood and details can be properly developed to optimize the performance of each type of insulation material. The specified requirements for each type of insulation that relate to moisture resistance are summarized in **table 3**. Without understanding the nuances of each test method used to determine water absorption or vapor adsorption, it appears that XPS and polyisocyanurate insulation are the most resistant to moisture, whereas mineral fiber and SPF insulation have the highest impact from moisture exposure, which is not necessarily the case.

For XPS insulation, testing of water absorption is required to follow ASTM C272, *Standard Test Method for Water Absorption of Core Materials for Sandwich Constructions*.[12] For this test, 12 in. by 12 in. by 1 in. test specimens are prepared and dried to get an initial dry weight. The samples are then submerged under a 1-in. head of water for 24 h. After the submersion period, the test specimen is "shaken vigorously," and all surfaces are wiped with a dry cloth until "no visible water is present." There are further provisions included if water is trapped on the surface to dip that portion of the specimen in isopropyl alcohol, shake the specimen again, and then allow the alcohol to evaporate. The specimen is then weighed to determine the wet weight, which is compared to the initial dry weight to determine the percent increase in weight, which is reported.

TABLE 3 Physical property requirements for moisture resistance for each type of insulation

	XPS (Type IV)	Mineral Fiber (Type IVB)	Polyisocyanurate (Type I)	SPF (Type II)
Test specimen size	12″ × 12″ × 1″	6″ × 6″ × full sample thickness	12″ × 12″ × 1″	6″ × 6″ × 3″
Test duration	24 h	96 h	2 h	96 h
Obtaining wet weight	Shake vigorously and wipe surfaces with dry cloth	Bag while in environmental chamber to lock in moisture	Drain for 10 min then blot with paper towel at each surface	Wet weight measured while submerged
Water absorption (ASTM C272)	0.3%	N/A	N/A	N/A
Water vapor sorption (ASTM C1104)	N/A	5%	N/A	N/A
Water absorption (ASTM C1763)	N/A	N/A	1%	N/A
Water absorption, volume (ASTM D2842)	N/A	N/A	N/A	5%

Polyisocyanurate insulation is tested in accordance with ASTM C209, *Standard Test Methods for Cellulosic Fiber Insulating Board*,[13] which then directs you to ASTM C1763, *Standard Test Method for Water Absorption by Immersion of Thermal Insulation Materials*, Procedure B.[14] The size of the test specimen required is not indicated within the material specification, but test specimens that are 12 in. by 12 in. are required in ASTM C1763, with the thickness of the test specimen to be representative of the manufactured product. Within the material specification, it is noted that a 1-in.-thick product was used to determine the physical property values outlined in the specification. The test specimens are preconditioned in accordance with ASTM C870 such that the specimen is placed in an oven at a temperature ranging from 212 to 248°F until moisture equilibrium is achieved, which essentially is the dry weight. The test specimen is then submerged under a 1-in. head of water for a period of 2 h. After submersion, the specimen is placed on end for 10 min to drain any excess water, and then it is permitted to lightly blot the specimen with a paper towel, "not to exceed two seconds per surface," prior to determining the weight following submersion. The presubmersion weight is then compared to the specimen weight after immersion, draining, and blotting to calculate the percent water absorbed by volume, which is reported. Within ASTM C209, a 24-h submersion test in accordance with Procedure C of ASTM C1763 can be performed if required, although this information is not readily available on manufacturer's product data sheets.

The water absorption of SPF insulation is determined in accordance with ASTM D2842, *Standard Test Method for Water Absorption of Rigid Cellular Plastics*.[15] Guidance on the size of the test specimen is not provided in the material specification, but it is noted in ASTM D2842 that test specimens are to be prepared that are 6 in. by 6 in. by 3 in. thick, unless materials cannot be manufactured to that thickness, in which case the thickness should be representative of the actual thickness for the material. After creating the test panels, they are required to cure for a minimum of 72 h at 73°F and 50% relative humidity prior to cutting to prepare samples for testing. No further conditioning requirements are outlined in the material specification, but in ASTM D2842, procedures are outlined to determine the dry weights by placing the insulation samples in an oven for 24 h to dry, followed by 4-h heating cycles. Two procedures for determining water absorption are outlined in ASTM D2842, and there is no indication which method is required in the material specification. For each procedure, the test specimens are submerged under a 2-in. head of water for a period of 96 h. The weight of the test specimens is taken while the specimen is submerged. For Procedure A, the water absorption is calculated using the dry weight and wet weight of the specimen following submersion. For Procedure B, an initial weight of the specimen is taken immediately following submersion and then compared to the final weight after the 96-h test period. The water absorption as a volume percent is then calculated and reported.

Mineral fiber insulation does not require testing of water absorption but rather evaluates water vapor adsorption in accordance with ASTM C1104, *Standard Test Method for Determining the Water Vapor Sorption of Unfaced Mineral Fiber Insulation*.[16] It is assumed that the mineral fiber insulation would be tested in accordance with Procedure A, which is designated for blanket, board, and pipe insulation products rather than Procedure B, which is for loose-fill insulation products. The size of the test specimens for mineral fiber are not outlined in the material specification, but a specimen size of 6 in. by 6 in., with the thickness being the "full sample" thickness, is indicated in ASTM C1104. The insulation is tested without any facings unless requested. Conditioning requirements are outlined in ASTM C1104 in which the dry weight of the test specimen is determined by placing the specimen in an oven for 2 h, followed by 10-min drying cycles until successive weight readings are consistent. The specimens are then heated to 140°F before placing them in an environmental chamber with ambient conditions of 120°F and a relative humidity of 95%. The specimens remain in the environmental chamber for 96 h, are bagged before removing them from the chamber, and then weighed. The weight of the specimen, minus the weight of the bag, is then compared with the initial dry weight of the specimen to determine the volume percent of water adsorbed, which is reported.

Each test method utilizes different test specimen sizes, different moisture exposure durations, and different methods for obtaining initial and final weights of the specimens. Although SPF insulation is submerged for the longest period of time, the surface area of the test specimen is less than half the surface area of the XPS insulation. The surface areas of XPS and polyisocyanurate insulation are comparable, but the duration in which they are submerged under water is vastly different. Whereas polyisocyanurate and XPS insulation are subjected only to a 1-in. head when submerged under water, and test specimens are weighed after they are removed from the water and allowed to drain and have excess surface water removed, SPF insulation is weighed while still submerged under a 2-in. head under water. With different test specimen sizes, the surface area exposed to moisture will vary. The differences in the duration the test specimens are exposed to moisture will impact how much moisture the material can absorb or adsorb within the test period. The methods for obtaining weights of the samples also causes discrepancies when comparing the results for each test method. Based on the material specifications for insulation, there is no clear way to compare the performance of these materials exposed to moisture.

VAPOR PERMEANCE

The vapor permeance of insulating materials does not necessarily have an impact on the performance of the material itself but is important information for the designer to understand so that all components of a wall assembly can be understood and evaluated to ensure the thermal and moisture gradients within the wall assembly result in favorable hygrothermal performance. XPS, polyisocyanurate, and SPF insulation are required to be tested in accordance with ASTM E96, *Standard Test Methods for Water Vapor Transmission of Materials*,[17] to determine the rate of water vapor transmission or vapor permeance. Although not specifically required by the mineral fiber material specification, many manufacturers still include the water vapor permeance on their product data sheets for reference. Two procedures that can be used to determine vapor permeance are outlined in ASTM E96: Desiccant Method and the Water Method. Test specimens are mounted and sealed to the top of a test dish that contains either desiccant or water. The specimen and test dish are weighed then placed in an environmental chamber held at a constant temperature of 73.4°F and 50% relative humidity. The specimens are weighed over a period of time, and the water vapor transmission rate is calculated from these data points, which are then used to determine the vapor permeance based on the vapor pressure difference, which is a factor of the temperature and relative humidity levels, across the test specimen.

For the Desiccant Method, the test specimen is exposed to a maximum relative humidity of 50% from the environmental chamber, as the relative humidity within the test dish is assumed to be between 0% and 3% from the desiccant. For the Water Method, the test specimen is exposed to a maximum relative humidity of 100% from the water within the test dish. If a material is hygroscopic, the exposure to different levels of relative humidity will result in a different vapor permeance. As such, the reported vapor permeance should be reflective of the end-use condition of the material. If a material is hygroscopic, the vapor permeance of the material when exposed to various relative humidity levels should be reported. All of the material specifications for insulation require test specimens to be tested in accordance with the Desiccant Method (**table 4**). For insulation materials installed within an exterior wall cavity, they will be exposed to relative humidity levels well above 50%. As such,

TABLE 4 Vapor permeance of each type of insulation

	XPS (Type IV)	Mineral Fiber (Type IVB)	Polyisocyanurate (Type I)	SPF (Type II)
Thickness	1 in.	N/A	1 in.	1 in.[a]
Facers	N/A	N/A	Yes	N/A
Test procedure	Desiccant method	N/A	Desiccant method	Desiccant method
Perms	1.5	N/A	0.3	3[a]

[a]Reported as water vapor permeability (perm-inches) in ASTM C1029. Value has been converted to permeance (perms) for a 1-in.-thick specimen for comparative purposes in this table.

consideration should be given to incorporating the Water Method into material specifications unless it can be proven that the material is nonhygroscopic and that the vapor permeance does not fluctuate at different relative humidity levels.

One key item noted with vapor permeance is it does vary with thickness. Generally, as the thickness of the material increases, the vapor permeance will decrease. Some materials will have a direct correlation where if the thickness of the material doubles then the permeance of the material is divided by two. Other materials may not correlate in this manner and will require additional coordination with individual manufacturers if the data are not reported on the product data sheet.

DIMENSIONAL STABILITY

The ability of thermal insulation to maintain its shape when exposed to a range of conditions is beneficial in that it will limit the gaps and voids between sections of insulation that may occur over time, which would create weak points in the building thermal envelope. These gaps and discontinuities become avenues for air leakage and water migration into or through the thermal barrier, which can reduce the in-service thermal resistance of the insulation material. Additionally, for insulation materials, such as SPF, that are adhered to the substrate and often other building enclosure accessories, dimensional stability can have an impact on continuity and performance of the air and moisture barrier systems of the exterior enclosure.

Most commonly, the test specimens are required to be tested in accordance with ASTM D2126, *Standard Test Method for Response of Rigid Cellular Plastics to Thermal and Humid Aging*,[18] for dimensional stability. This test method is intended for use for rigid cellular plastics, which excludes mineral fiber insulation from being tested in accordance with this standard. For this test, the length, width, and thickness of test specimens is carefully measured, and then the samples are placed in an environmental chamber at a constant temperature and relative humidity specified by the material specification. Although unique conditions are included in each material specification, requiring the test specimens to be exposed to a temperature of 158°F and 97% relative humidity is included in all of the material specifications. Per ASTM D2126, samples are exposed to the specified temperature and relative humidity combinations for 12 h, 7 days, and 14 days; however, the material standards for XPS, polyisocyanurate, and SPF insulation materials specify dimensional stability based on measurements taken at 7 days. After each exposure, specimens are allowed to cool to room temperature for 2 h, and the dimensions are remeasured such that changes in dimensions can be calculated and reported. Mineral fiber insulation is tested in accordance with ASTM C356, *Standard Test Method for Linear Shrinkage of Preformed High-Temperature Thermal Insulation Subjected to Soaking Heat*.[19] Similar procedures are followed, except the specimen is exposed to high temperatures for only 24 h. The specified test methods and supplemental information relating to each test method are outlined in **table 5**.

Although most of the insulation materials are tested in accordance with ASTM D2126, the size of the test specimen for each material type varies. Because dimensional change as determined by ASTM D2126 and ASTM C356 are reported in terms of percent change, the length of change is relative to the original sample size. Within the material specification for XPS insulation, a size for the test specimen is not clearly specified other than noting it should be 1 in. thick. The indicated dimensions for polyisocyanurate and SPF insulation are also different. For each type of insulation, test conditions of 158°F and 97% relative humidity are outlined, with requirements for other test conditions that include low temperature exposure in some of the material specifications. The reporting of the dimensional stability also varies by orientation of the specimen. For polyisocyanurate insulation, the linear shrinkage is reported based on the thickness of the insulation from the other directions, whereas XPS and SPF insulation account for dimensional change in any direction. For mineral fiber insulation, the dimensional change must be reported only in the longest direction of the test specimen.

TABLE 5 Physical properties for dimensional stability for each type of insulation

	XPS (Type IV)	Mineral Fiber (Type IVB)	Polyisocyanurate (Type I)	SPF (Type II)
Test method	D2126	C356	D2126	D2126
Test specimen size	1" thick	6" × 12" × single layer	12" × 12" × 1"	12" × 12" × 1½"
Conditioning	73.4°F, 50% RH	Dry prior to testing	Per D2126, 73.4°F and 50% RH	Per D2126, 73.4°F and 50% RH
Test conditions	158°F, 97% RH; −40°F, ambient RH	Max. use temperature (Max. 300°F)	158°F, 97% RH; −40°F, ambient RH; 200°F, ambient RH	158°F, 97% RH
Test duration	7 days	24 hours	7 days	7 days
Change in dimensions	2%	2% (12" dimension)	4% (thickness, 158°F); 2% (length/width, 158°F)	9% in any direction

Although the test methods used to determine dimensional stability are nearly the same for each type of insulation, variations in the specimen size and reporting requirements do limit the ability of design professionals to accurately compare these materials. Furthermore, the impacts of exposure at cold temperatures are not consistently required for each material, and it is unclear whether the change reported is in expansion or shrinkage, except for mineral fiber insulation that indicated the change in linear shrinkage. Additionally, it should be noted that the tests for dimensional stability are intended to test dimensional changes at in-service conditions and ignore dimensional changes that might occur in the fabrication process. For many materials, initial dimensional changes associated with fabrication do not impact installed conditions; however, for products such as SPF, which are manufactured on site, these early dimensional changes should not be ignored. Consideration should be given to utilizing the same test specimen size for each type of insulation material and reporting the dimensional change as shrinkage or expansion in each direction after testing. It is unclear why the testing requirements for mineral fiber insulation vary so greatly from the rigid cellular types of insulation materials, but this creates even greater disparity in being able to compare the physical properties of these materials.

STRENGTH

When installed in an exterior wall cavity, insulation is not intended to transfer any loads unless it is designed to serve as part of the structural system for the building; therefore, the strength properties are generally not as essential for the design of the wall assembly. In roofing applications, the compressive strength of insulation is an important consideration such that the insulation can resist the anticipated live loads on the roof structure. Some type of compression-related property, whether it is compressive strength or compressive resistance, are included in the material specifications for each type of insulation. Flexural strength or tensile strength are included in the material specifications for some of the insulation types. In some designs, the insulation is intended to serve as part of the structural system of the building, in which case understanding these material properties becomes critically important to the stability of the building. The strength requirements for each type of insulation are outlined in **table 6**.

When evaluating compression, either the compressive strength (ASTM D1621, *Standard Test Method for Compressive Properties of Rigid Cellular Plastics*[20]) or compressive resistance (ASTM C165, *Standard Test Method for Measuring Compressive Properties of Thermal Insulations*[21]) are noted in the material specification, or sometimes both. Each type of test requires the insulation to be mounted in a compression machine, with loading gradually applied until the failure modes are met. Failure modes are considered to be when the test specimen reaches 10% deformation or at yield, whichever occurs first. The main difference in the test methods is the loading mechanism from which the load is applied. Within ASTM C165, the loading mechanism incorporates a spherical bearing block assembly that is

TABLE 6 Strength requirements for each type of insulation

	XPS (Type IV)	Mineral Fiber (Type IVB)	Polyisocyanurate (Type I)	SPF (Type II)
Compressive strength (ASTM D1621)	Permitted, requirements not reported	N/A	16 psi Procedure not specified	25 psi Procedure A
Compressive resistance (ASTM C165)	25 psi Procedure A	Not required (Category 1) 50 psf (Category 2) Procedure not specified	N/A	Permitted, requirement not reported Procedure A
Flexural strength (ASTM C203) Method I, Procedure B	50 psi	N/A	40 psi min; 8 lb. min. break load	N/A
Tensile strength (ASTM C209)	N/A	N/A	500 psf	N/A
Tensile strength (ASTM D1623)	N/A	N/A	N/A	32 psi

suspended from the machine and not mounted rigidly. This allows for lateral movement in the insulation material as compressive forces are being applied perpendicular to the surface of the insulation. Within ASTM D1621, the loading mechanism is rigid. Either test method is permitted in the material specifications for XPS and SPF insulation to evaluate compression of the insulation, although in the table contained within the specification that outlines physical property requirements, the results appear to be provided for only one of the test methods, as designated by the terminology of "strength" or "resistance," which makes it appear the test results are interchangeable even though the test methods can produce different results. The test specimen size for each test method is not clearly defined in each of the material specifications, which could impact the results of the test or alter the failure modes that get reported.

Flexural strength requirements are only outlined in the material specification for XPS and polyisocyanurate insulation. This property is generally associated with a material's ability to span between supports, which would be applicable if the insulating material were to be installed as sheathing where it spans between stud framing. Tensile strength is required for polyisocyanurate insulation and SPF insulation, although the test methods specified measure very different properties. The tensile strength for polyisocyanurate insulation (ASTM C209) is measured perpendicular to the major surface of the board and only for faced board products. Essentially, this will measure the adhesion of the facer to the polyisocyanurate foam plastic. The tensile strength for SPF insulation aims to measure the internal tensile strength of the material (ASTM D1623, *Standard Test Method for Tensile and Tensile Adhesion Properties of Rigid Cellular Plastics*[22]).

OTHER MATERIAL PROPERTIES

In addition to the material properties used to classify each type of insulation, test methods to determine other physical properties for each type of material are included in the material specification as general requirements. Within an exterior cavity wall application, the fire resistance of the materials that are installed must be considered when designing a wall assembly. When materials in the exterior wall assembly are combustible, compliance with NFPA 285 must then be considered when designing the overall assembly. The wet, humid climate that can exist within an exterior wall assembly is also conducive to fungal growth. The ability of materials installed in the exterior wall cavity to resist fungal growth can impact the indoor air quality of the building. Material compatibility within the exterior wall assembly is critical to ensure the long-term performance of the assembly. When cladding is installed with steel components that penetrate through the insulation layer, understanding the potential chemical reactions that can occur that could corrode these steel elements is important. From an occupant comfort standpoint, the ability of thermal insulation to not only provide thermal comfort but also reduce exterior noise can also prove to be a valuable characteristic.

Fire Resistance

Of the insulation types studied in this paper, mineral fiber insulation is the only noncombustible insulation type. The surface burning characteristics are part of the classification of this material, including testing to determine the flame spread index and smoke development. These properties are determined in accordance with ASTM E84, *Standard Test Method for Surface Burning Characteristics of Building Materials*.[23] The other insulation materials reference this test method if the surface burning characteristics are desired by the end user, but they are only required to be reported and do not need to meet certain requirements to classify the combustible materials. Generally, manufacturers will report these characteristics on product data sheets such that the information is readily available to the end user and to show compliance with building code requirements for surface burning characteristics of foam plastic insulation.

Fungi Resistance

Fungi resistance is tested for in accordance with ASTM C1338, *Standard Test Method for Determining Fungi Resistance of Insulation Materials and Facings*,[24] and is required within the material specification for mineral fiber insulation. For this test, fungal spores are introduced to test specimens that are placed in a controlled environment to evaluate the extent to which growth occurs over an incubation period. Although not required by all material specifications, manufacturers of other insulating materials may report the material's ability to resist mold growth through this test method or other test methods. Standardization of the approach to evaluating mold resistance on insulation products should be considered such that the material specifications each reference the same test standard.

Corrosiveness to Steel

A test method to evaluate the corrosiveness of various metals when exposed to the mineral fiber is included within the material specification for mineral fiber insulation. Procedures are outlined in ASTM C665, *Standard Specification for Mineral-Fiber Blanket Thermal Insulation for Light Frame Construction and Manufactured Housing*,[25] for testing steel, copper, and aluminum plates that are sandwiched between samples of mineral fiber insulation and suspended in an environmental chamber at a constant temperature of 120°F and 95% relative humidity for an extended period of time. Procedures are outlined in ASTM C795, *Standard Specification for Thermal Insulation for Use in Contact with Austenitic Stainless Steel*,[26] for evaluating external stress corrosion cracking in stainless steel test specimens when exposed to insulation and bulk water over an extended period of time as well as providing limits for chemicals within insulation that are known to be corrosive to stainless steel.

It is important that the chemicals comprising various types of insulating materials do not corrode steel elements within an exterior wall assembly, such as cladding attachments, shelf angles, or flashing. This could lead to catastrophic failures in the structural elements of the wall system. As such, consideration should be given to incorporating ASTM C665 and ASTM C795 into each material specification such that insulating materials can be compared uniformly as it relates to this physical property.

Acoustical Performance

Acoustical performance is not required to be reported in any of the material specifications for insulation; however, many manufacturers report these properties on product data sheets. Because of the lack of standardization, a variety of test methods are used to evaluate acoustical performance, which makes it difficult to equally compare each type of product. Several of the acoustical performance test methods identified in manufacturers' product data sheets are ASTM C423, *Standard Test Method for Sound Absorption and Sound Absorption Coefficients by the Reverberation Room Method*,[27] ASTM E90, *Standard Test Method for Laboratory Measurement of Airborne Sound Transmission Loss of Building Partitions and Elements*,[28] and ASTM E2179, *Standard Test Method for Laboratory Measurement of the Effectiveness of Floor Coverings in Reducing Impact Sound Transmission through Concrete Floors*.[29] A reverberation room is utilized in ASTM C423 in which various frequency sounds are made within the room when the room is empty and when a test specimen is then placed within the room. The differences in the sounds are then used to calculate the sound absorption coefficient at each frequency, which can then be converted in a rating on the noise reduction coefficient scale from 0 to 1, where either no sound is absorbed or all sound is absorbed, respectively. The reduction in sound transmission is tested in accordance with ASTM E90, in which a test specimen is mounted within a partition wall between two adjacent rooms and a sound is created in one room and the reduction in sound transmission is measured in the other room. The sound transmission loss can be classified by the sound transmission class or outdoor-indoor transmission class. The reduction in sound transmission through a concrete floor slab when the test specimen is introduced into the assembly is determined using ASTM E2179. This test method is used to determine the impact insulation class. Each of these acoustical ratings provides different information and are more applicable in different applications. For insulation used in an exterior wall assembly, the use of ASTM E90 is the most applicable because this standard evaluates the reduction in sound between two spaces, which is how you would expect the exterior wall assembly to perform.

EXCLUDED MATERIAL PROPERTIES

In addition to the properties included in the ASTM standard specification for each insulation type discussed in this paper, there are other notable properties that often play a key role in the selection of materials by design professionals. Although the primary function of thermal insulation materials is to form the thermal barrier for the building enclosure, design professionals are increasingly exploring opportunities for thermal insulation to act as a component of the air and water barrier for exterior wall assemblies. Many manufacturers of XPS, polyisocyanurate, and SPF insulation materials offer published literature indicating the acceptable use of insulating materials as air and weather barriers in exterior wall applications. However, the nonstandardization of these properties leads to variations in how they are reported and challenges when comparing the anticipated performance of different insulation materials as air and weather barriers.

Water-Resistive Barrier

The requirements for water resistive barriers are established in Chapter 14 of the International Building Code. According to the code, a water resistive barrier must comply with ASTM D226,[30] which is a material specification for building felt, or "other approved materials." Not all insulation types are intended to serve as a water-resistive barrier; generally, XPS, foil-faced polyisocyanurate insulation, and closed-cell SPF insulation are the only types of insulation that could also be considered water-resistive barriers. Typically, design professionals can ascertain the ability of an insulation material to act as a water-resistive barrier based on the technical data sheets published by the manufacturers. Many of these manufacturers provide product literature that markets the insulation material for exterior weather-resistive barrier applications, but often the design professional must review the ICC-ES reports for the material to understand the specific requirements for weather-barrier applications. In many cases, variability exists between similar insulation materials supplied by different manufacturers, which can create confusion when specifying the requirements for insulation materials intended to serve as a weather-resistive barrier. As an example, some closed-cell SPF products must be installed at a minimum thickness of 1 in., whereas others must be installed at a minimum thickness of $1\frac{1}{2}$ in. Similarly, valuable information is provided in ICC-ES reports for the use of board stock products, such as XPS and foil-faced polyisocyanurate, as weather-resistive barriers; however, the specific installation instructions for each material, including acceptable types of fasteners, joint sealing tape, and penetration sealing requirements, vary between manufactured products.

Although the primary function of the insulation is to serve as a thermal layer, the determination of whether the insulation can or will function as a water-resistive barrier can impact design assumptions. Hygrothermal models have become a common design tool to assess the location of materials and material properties in the typical opaque assemblies of the building enclosure to minimize the potential for long-term moisture-related issues. These models are typically developed in accordance with ASHRAE 160,[31] which states that, in the absence of full-scale testing, 1% of the water reaching the exterior surface of the assembly shall be placed at the exterior surface of the water-resistive barrier. Consequently, the determination of whether an insulation layer is a water-resistive barrier for a specific project can impact the placement of these moisture sources within the material layers of the model and, therefore, greatly affect the predicted performance. Additionally, the design professional must accurately ascertain the locations within an assembly where water is to be managed to detail and specify drainage provisions appropriately.

Air Barrier

The ability of a material to function as an air barrier is determined in accordance with ASTM E2178[32] as outlined in the International Energy Conservation Code. For this test, a material sample is covered with a polyethylene sheet, creating an air seal, and mounted and sealed within a test chamber. The air leakage through the test chamber is measured with and without the polyethylene sheet intact at various static pressure differentials across the test specimen. The air permeance of the material can then be determined on the basis of the measured air leakage rates and pressure differentials. Similar to the weather-resistive barriers, not all types of insulation are intended to serve as the air barrier within the building thermal envelope. Even with defined guidance on how to determine the air permeance of a material, the material specifications for insulation do not reference these requirements; however, manufacturers that intend for their insulation products to serve as air barriers do report the air permeance of the material in accordance with ASTM E2178.

Although the field of the insulation material may meet the requirements of an air barrier, the installation requirement will dictate the ability of the insulation material to serve as a continuous air barrier. These requirements are typically defined most comprehensively in ICC-ES reports for each individual insulation product. There are variations in material installation requirements for different products that must be followed to ensure the insulation material can function as an air barrier, which can make specifying insulation as a continuous air barrier challenging for the design professional to capture the broad range of installation requirements.

Even when the insulating material is not intended to serve as the air barrier for the exterior enclosure, the impact of air movement on the insulation performance should be considered. The ability of air to move through an insulation layer or between discrete panels of insulation can impact the in-service thermal resistance of the exterior enclosure.

Discussion

After performing a detailed review of the material specifications for various types of insulation, it becomes clear that there are very few similarities between how each type of insulation material is required to be tested. For design professionals, information that is reported on a manufacturer's product data sheet will generally guide decisions that are made when specifying insulation products within an exterior wall assembly. Without understanding the nuances of the test methods and, therefore, not realizing the variations in the physical properties between materials, the potential exists for insulation materials to be specified and installed within a wall assembly that will not perform as intended. Although it may be difficult to incorporate the same test methods into each material specification, consensus is needed on which physical properties and corresponding test methods will capture the actual performance of insulation materials when installed in exterior cavity wall applications, and this information should be made readily available to design professionals.

As the industry continues to push for higher-performing buildings, the assessment of an insulation material based on thermal performance alone has already become a dated practice. A full understanding of the material properties of these products is becoming more critical to understand to evaluate the building envelope holistically. The most notable discrepancy in the specified test methods was related to the evaluation of the moisture resistance of the materials. In cases in which insulation is installed outboard of the water management layer of the assembly, it will consistently be exposed to bulk water. The current information provided to design professionals does not provide clear information that would allow proper detailing to optimize the performance of each type of insulation when installed in the exterior wall cavity. Furthermore, when installed in the exterior wall cavity, it is expected that the insulation will also be subjected to high levels of water vapor. Understanding the impacts of the exposure to moisture on the thermal performance of the insulation, which is the primary function of the insulation, is necessary to ensure the energy modeling and simulations that are performed to evaluate buildings are representative of the actual built conditions. As a starting point, consideration should be given to testing the thermal performance of the insulation material immediately following the moisture resistance testing and testing insulation materials in accordance with ASTM C1512. A more detailed and accurate representation of the thermal performance of the insulating materials would be to test the thermal resistance after exposing the test specimens to different levels of moisture through conditioning.

In an effort to simplify designs and improve the efficiency of construction, some insulation products are marketed as air barriers, water management layers, or even structural components of the building envelope in addition to functioning as the thermal resistance layer. Additionally, design professionals are increasingly more reliant on a comprehensive understanding of material properties to properly design the exterior enclosure. Due to the fact that not all insulation materials can be installed to serve as a water barrier, air barrier, or structural component, we appreciate the fact that these material properties likely should not be incorporated in the material standards, as this could inadvertently exclude materials that solely act as a thermal barrier. However, consideration should be given to the inclusion of an appendix or the creation of a standard guide that provides the industry with a more standardized method for evaluating the impacts of these properties on the insulation and how to specify insulation materials to be used as air and water barriers.

Last, the sustainability of products used in building construction is also becoming more important to building owners when selecting materials. Consideration should be given to expanding test methods and performance requirements beyond just the in situ performance and evaluating the life-cycle costs or environmental impact from the manufacturing processes and chemical composition of each insulation material.

References

1. *Standard Specification for Rigid, Cellular Polystyrene Thermal Insulation*, ASTM C578-19 (West Conshohocken, PA: ASTM International, approved June 1, 2019), http://doi.org/10.1520/C0578-19
2. *Standard Practice for Sampling and Acceptance of Thermal Insulation Lots*, ASTM C390-08(2019) (West Conshohocken, PA: ASTM International, approved September 1, 2019), http://doi.org/10.1520/C0390-08R19
3. *Standard Practice for Conditioning of Thermal Insulating Materials*, ASTM C870-11(2017) (West Conshohocken, PA: ASTM International, approved April 15, 2017), http://doi.org/10.1520/C0870-11R17
4. *Standard Specification for Mineral Fiber Block and Board Thermal Insulation*, ASTM C612-14(2019) (West Conshohocken, PA: ASTM International, approved September 1, 2019), http://doi.org/10.1520/C0612-14R19
5. *Standard Specification for Flexible, Low Permeance Vapor Retarders for Thermal Insulation*, ASTM C1136-17A (West Conshohocken, PA: ASTM International, approved October 15, 2017), http://doi.org/10.1520/C1136-17A
6. *Standard Specification for Faced Rigid Cellular Polyisocyanurate Thermal Insulation Board*, ASTM C1289-20 (West Conshohocken, PA: ASTM International, approved October 1, 2020), http://doi.org/10.1520/C1289-20
7. *Standard Specification for Spray-Applied Rigid Cellular Polyurethane Thermal Insulation*, ASTM C1029-20 (West Conshohocken, PA: ASTM International, approved December 1, 2020), http://doi.org/10.1520/C1029-20
8. *Standard Test Method for Steady-State Thermal Transmission Properties by Means of the Heat Flow Meter Apparatus*, ASTM C518-17 (West Conshohocken, PA: ASTM International, approved May 1, 2017), http://doi.org/10.1520/C0518-17

9. *Standard Practice for Selecting Temperatures for Evaluating and Reporting Thermal Properties of Thermal Insulation*, ASTM C1058/C1058M-10(2015) (West Conshohocken, PA: ASTM International, approved March 1, 2015), http://doi.org/10.1520/C1058_C1058M-10R15
10. *Standard Test Method for Thermal Performance of Building Materials and Envelope Assemblies by Means of a Hot Box Apparatus*, ASTM C1363-19 (West Conshohocken, PA: ASTM International, approved September 1, 2019), http://doi.org/10.1520/C1363-19
11. *Standard Test Method for Characterizing the Effect of Exposure to Environmental Cycling on Thermal Performance of Insulation Products*, ASTM C1512-10(2020) (West Conshohocken, PA: ASTM International, approved June 15, 2020), http://doi.org/10.1520/C1512-10R20
12. *Standard Test Method for Water Absorption of Core Materials for Sandwich Constructions*, ASTM C272/C272M-18 (West Conshohocken, PA: ASTM International, approved August 1, 2018), http://doi.org/10.1520/C0272_C0272M-18
13. *Standard Test Methods for Cellulosic Fiber Insulating Board*, ASTM C209-20 (West Conshohocken, PA: ASTM International, approved March 15, 2020), http://doi.org/10.1520/C0209-20
14. *Standard Test Method for Water Absorption by Immersion of Thermal Insulation Materials*, ASTM C1763-20 (West Conshohocken, PA: ASTM International, approved May 1, 2020), http://doi.org/10.1520/C1763-20
15. *Standard Test Method for Water Absorption of Rigid Cellular Plastics*, ASTM D2842-19 (West Conshohocken, PA: ASTM International, approved May 1, 2019), http://doi.org/10.1520/D2842-19
16. *Standard Test Method for Determining the Water Vapor Sorption of Unfaced Mineral Fiber Insulation*, ASTM C1104/C1104M-19 (West Conshohocken, PA: ASTM International, approved September 1, 2019), http://doi.org/10.1520/C1104_C1104M-19
17. *Standard Test Methods for Water Vapor Transmission of Materials*, ASTM E96/E96M-16 (West Conshohocken, PA: ASTM International, approved March 1, 2016), http://doi.org/10.1520/E0096_E0096M-16
18. *Standard Test Method for Response of Rigid Cellular Plastics to Thermal and Humid Aging*, ASTM D2126-20 (West Conshohocken, PA: ASTM International, approved December 1, 2020), http://doi.org/10.1520/D2126-20
19. *Standard Test Method for Linear Shrinkage of Preformed High-Temperature Thermal Insulation Subjected to Soaking Heat*, ASTM C356-17 (West Conshohocken, PA: ASTM International, approved May 1, 2017), http://doi.org/10.1520/C0356-17
20. *Standard Test Method for Compressive Properties of Rigid Cellular Plastics*, ASTM D1621-16 (West Conshohocken, PA: ASTM International, approved May 1, 2016), http://doi.org/10.1520/D1621-16
21. *Standard Test Method for Measuring Compressive Properties of Thermal Insulations*, ASTM C165-07(2017) (West Conshohocken, PA: ASTM International, approved September 1, 2017), http://doi.org/10.1520/C0165-07R17
22. *Standard Test Method for Tensile and Tensile Adhesion Properties of Rigid Cellular Plastics*, ASTM D1623-17 (West Conshohocken, PA: ASTM International, approved May 1, 2017), http://doi.org/10.1520/D1623-17
23. *Standard Test Method for Surface Burning Characteristics of Building Materials*, ASTM E84-20 (West Conshohocken, PA: ASTM International, approved March 1, 2020), http://doi.org/10.1520/E0084-20
24. *Standard Test Method for Determining Fungi Resistance of Insulation Materials and Facings*, ASTM C1338-19 (West Conshohocken, PA: ASTM International, approved March 1, 2019), http://doi.org/10.1520/C1338-19

25. *Standard Specification for Mineral-Fiber Blanket Thermal Insulation for Light Frame Construction and Manufactured Housing*, ASTM C665-17 (West Conshohocken, PA: ASTM International, approved September 1, 2017), http://doi.org/10.1520/C0665-17
26. *Standard Specification for Thermal Insulation for Use in Contact with Austenitic Stainless Steel*, ASTM C795-08(2018) (West Conshohocken, PA: ASTM International, approved November 1, 2018), http://doi.org/10.1520/C0795-08R18
27. *Standard Test Method for Sound Absorption and Sound Absorption Coefficients by the Reverberation Room Method*, ASTM C423-17 (West Conshohocken, PA: ASTM International, approved January 15, 2017), http://doi.org/10.1520/C0423-17
28. *Standard Test Method for Laboratory Measurement of Airborne Sound Transmission Loss of Building Partitions and Elements*, ASTM E90-09(2016) (West Conshohocken, PA: ASTM International, approved December 1, 2016), http://doi.org/10.1520/E0090-09R16
29. *Standard Test Method for Laboratory Measurement of the Effectiveness of Floor Coverings in Reducing Impact Sound Transmission through Concrete Floors*, ASTM E2179-03(2016) (West Conshohocken, PA: ASTM International, approved April 1, 2016), http://doi.org/10.1520/E2179-03R16
30. *Standard Specification for Asphalt-Saturated Organic Felt Used in Roofing and Waterproofing*, ASTM D226/D226M-17 (West Conshohocken, PA: ASTM International, approved June 15, 2017), http://doi.org/10.1520/D0226_D0226M-17
31. *Criteria for Moisture-Control Analysis in Buildings*, ASHRAE 160 (Atlanta, GA: ASHRAE, 2016).
32. *Standard Test Method for Determining Air Leakage Rate and Calculation of Air Permeance of Building Materials*, ASTM E2178-21 (West Conshohocken, PA: ASTM International, approved February 1, 2021), http://doi.org/10.1520/E2178-21

STP 1635, 2022 / available online at www.astm.org / doi: 10.1520/STP163520200115

Martina T. Driscoll,[1] Andrea DelGiudice,[1] and Adrienne Larson[2]

Pillow Talk: ETFE Roofing Design and Quality-Control Testing

Citation
M. T. Driscoll, A. DelGiudice, and A. Larson, "Pillow Talk: ETFE Roofing Design and Quality-Control Testing," in *Building Science and the Physics of Building Enclosure Performance: 2nd Volume*, ed. D. J. Lemieux and J. Keegan (West Conshohocken, PA: ASTM International, 2022), 53–70. http://doi.org/10.1520/STP163520200115[3]

ABSTRACT
Ethelene tetrafluoroethylene (ETFE) assembly is utilized on thousands of structures worldwide in both wall and roofs/skylight applications. It is used in multiple configurations, including the "pillow" configuration explored in this paper, and was installed at the Beijing National Aquatics Center (2008) (nicknamed the Watercube), the Eden Project in the United Kingdom (2001), Ferrari World in Abu Dhabi (2010), and U.S. Bank Stadium (home of the Vikings) in Minnesota (2016). Although the ETFE assembly can be used in lieu of traditional glazing and can be compared to traditional curtain wall or skylight assemblies in several ways, it also varies significantly from traditional glass glazing. Namely, it is a fraction of the weight, is transparent to ultraviolet and long-wave radiation (unless coated to perform otherwise), and responds to structural loads much differently than glass. This paper provides an overview of the ETFE pillow assembly, explores design and quality-assurance considerations for the assembly during construction, and draws on two case studies to demonstrate testing methodologies related to bulk water penetration: structural performance and deflection of the pillow assembly.

Keywords
ethylene tetrafluoroethylene, ETFE, structural, testing, quality control, quality assurance

---

Manuscript received December 11, 2020; accepted for publication August 5, 2021.
[1] Wiss, Janney, Elstner Associates, Inc., 2941 Fairview Park Dr., Ste. 300, Falls Church, VA 22032, USA M. T. D. http://orcid.org/0000-0002-0364-3883, A. D. https://orcid.org/0000-0002-8623-3774
[2] Wiss, Janney, Elstner Associates, Inc, 3609 S. Wadsworth Blvd., Ste. 400, Lakewood, CO 80235, USA http://orcid.org/0000-0002-3485-611X
[3] ASTM Second Symposium on *Building Science and the Physics of Building Enclosure Performance* on April 24–25, 2022, and June 12, 2022 in Seattle, WA, USA.

Copyright © 2022 by ASTM International, 100 Barr Harbor Drive, PO Box C700, West Conshohocken, PA 19428-2959.

ASTM International is not responsible, as a body, for the statements and opinions expressed in this paper. ASTM International does not endorse any products represented in this paper.

Background

Ethylene tetrafluoroethylene (ETFE) is a modified copolymer developed by DuPont for NASA use due to its corrosion resistance and strength over a high temperature range. It is described as a thermoplastic version of Teflon and is used in multiple applications in the industry.

Dr. Stefan Lehnert, then a mechanical engineering student, originally investigated ETFE when he was looking for better foils for his sailboat. He noted ETFE to be transparent, self-cleaning, durable, and flexible at long lengths. He also recognized its potential use as a building material and, in 1982, founded Vector Foiltec (Foiltec) in Bremen, Germany. Foiltec is the design and manufacturing company that turned ETFE into the "Texlon Foil System" for use as part of the building enclosure.[1] Following Foiltec, traditional "tensile roof" manufacturers, including Birdair, Architen Landrell, Skyspan, and Base Structures Ltd., have adopted the material and system concept for various applications in lieu of other "fabrics" such as Teflon-coated fiberglass (used at the Denver International Airport, among other buildings).

ETFE is currently being utilized in thousands of structures worldwide in both wall and roof/skylight applications. Notable projects utilizing ETFE include the 2008 Beijing National Aquatics Center (nicknamed the Watercube), the Eden Project in the United Kingdom (2001), Ferrari World in Abu Dhabi (2010), and U.S. Bank Stadium (home of the Vikings) in Minnesota (2016).[2]

System Description

Assemblies constructed of ETFE are lightweight, transparent, have thermal properties comparable to or better than glass, and are shatterproof. However, if optical clarity, ultraviolet (UV) resistance, and acoustics are concerns, ETFE may be a less desirable choice. This paper discusses the use of ETFE systems as a building enclosure system and presents two case studies that highlight performance and installation considerations as well as a rational quality-control and performance-testing approach.

When ETFE is used in a skylight or curtain wall application, the copolymer is extruded into thin films (termed by Foiltec as "foils"—a nod to its nautical history) that can be used to form either a single-layer membrane or multilayer pillows (typically up to five layers). The pillows are then filled with air and "pneumatically pre-stressed."[3] **Figure 1** is a conceptual sketch showing a three-layer pillow system glazed into a typical extrusion. The films are sealed to each other and to a strip of foil folded over a "keder" rod or rope, which is supported in an aluminum perimeter extrusion (typically referred to as a "keder track"), which, in turn, is supported by the underlying building structure.[3,4]

The pillows are kept continually pressurized by a small air handling unit (AHU) that maintains the pressure at approximately 200 Pa. This pressurization provides the pillows with structural stability and some insulative properties.[5]

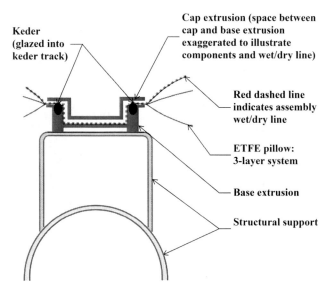

FIG. 1 Sketch showing the keder retained in the frame extrusion, which is attached to structural framing beneath; the system's wet/dry line is shown in bold and dashed lines.

Although some systems may incorporate more than one AHU unit due to size or desired redundancy, a large skylight can be powered by a single AHU that contains two fans powered by electric motors. Dehumidifiers are typically used in conjunction with the blowers to minimize the risk for condensation within the pillows. The system requires a dedicated and secure power supply. The air pressure in the pillows can be increased to resist higher loads, which can be controlled through sensors in the system.

Should a power failure occur, the pillows can maintain inflation for between 3 and 6 h (depending on weather conditions). Therefore, an emergency generator is typically recommended.

System Properties

SIZE AND WEIGHT

The manufacturing process limits the width of the foil to a maximum of 1.5 m, although broader sheets can be produced by heat-welding foils together. The manufacturer's data indicate that rectangular pillows can span up to 3.5 m in the short direction with no stated limitations for the long direction (see **fig. 2** for an example of rectangular pillows). For triangular pillows, the size can reportedly be increased.[5] ETFE foil pillows typically weigh 2–3.5 kg/m^2, less than 2% of the weight of glass,

FIG. 2 View of foil pillow installation in progress.

whereas the entire ETFE system, including the aluminum and steel support structure, weighs between 10% and 50% of conventional glass assemblies.[3]

Transparency

Typically, the visible light transmission of clear foils is approximately 95%. The thermal and optical properties of the ETFE pillows can be altered significantly by the application of coatings, print, geometry, and the layering in which they are applied. As shown in **figure 3**, a reflective frit can also be applied to an inflatable intermediate pillow, and the intermediate foils can be in an open or closed position to allow or restrict variable amounts of daylight (and heat) into the inner space, and coatings can be applied to reduce UV transparency.[6]

FIG. 3 Diagram of dynamic shading using frit-pattern, reflective frit at intermediate layers.[6]

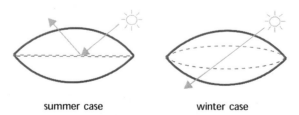

summer case   winter case

Thermal Performance

Table 1 compares the U-values (thermal transmittance) of glass and ETFE. Note that the g-value (total solar energy transmittance) is equivalent to the solar heat gain coefficient (ratio of solar heat gain through the material relative to the incident solar radiation on the glazing). ETFE is reported to have a better insulation value than triple glazing when used horizontally. (Note that glazing manufacturer's figures are for vertical glazing, which increases those numbers.[6])

In traditional insulating glass unit (IGU) design, the application of a low-emissivity coating will reduce the emissivity of one of the panes and reduce the long-wave radiation exchange and thus the thermal transmittance of the system. By filling the IGU cavities with argon or krypton, the heat exchange due to convection between the panes is reduced, achieving similar results (lower U-values). Similarly, in ETFE, coatings can be applied to control heat gain and loss. A low-emissivity or solar control coating, or both, can be applied to reduce the long-wave transmission losses (i.e., during a cold winter night providing lower thermal transmittance values) or the solar transmittance, respectively, as shown in **figure 4**. In contrast to IGUs, because the pillows are inflated with air continually managed by AHUs, argon or krypton fill would not be practical.[6]

As described by Poirazis et al.,[6] ETFE pillows typically incorporate two or three air chambers formed by layering foils. Convective heat transfer within these air

TABLE 1 Glass glazing and ETFE U- and g-values[6]

	U-value (W/m²K)	g-value
6-mm monolithic glass	5.9	0.95
6-12-6-mm double glazing unit	2.8	0.83
6-12-6-mm high-performance double glazing unit	2.0	0.35
Two-layer ETFE pillow	2.9	0.71–0.22[a]
Three-layer ETFE pillow	1.9	0.71–0.22[a]
Four-layer ETFE pillow	1.4	0.71–0.22[a]

[a]With frit.

FIG. 4 Diagram showing low-emissivity and solar control coating effect on solar and long-wave radiation.[6]

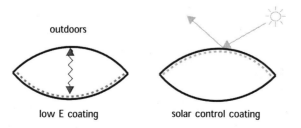

chambers will influence the thermal performance of the pillow. Therefore, despite the values provided in table 1, the estimation of accurate U-values is generally complex. Modeling of ETFE pillows as part of building performance simulation is not straightforward. The performance of the system at full inflation can be assessed by means of computational fluid dynamics or empirical (hot box) testing, or both.[6] However, the system inflation level is dynamic and affected by the current loading (wind, snow, etc.). Pillow inflation or deflation due to changes in ambient temperature (both seasonal and diurnal) are unlikely to have a significant effect because these changes are quickly corrected by the AHU.

Acoustics

ETFE assemblies have been termed "acoustically transparent." Although this can be advantageous for a stadium roof ("letting noise out"), the pillows can also "let noise in." Noise levels up to 80 dB have been observed due to events such as rain droplets impacting the pillows and creating a "drumming" noise. The manufacturers state that rain noise can be suppressed using a rain attenuation/dampening layer "mesh" added to the top surface of the pillows, minimizing the sound reverberating in the space below. The relative effectiveness of this attenuation layer has been demonstrated through research performed in the Netherlands.[7]

Additional Considerations

Foiltec has developed a smoke-venting system for their ETFE roofs, in which a "hot wire" carrying a current linked to an alarm cuts the foil and lets it drop down to provide smoke evacuation. Sprinklers can also be installed to the system. Foiltec has also patented a method of laminating photovoltaic cells into ETFE and has built the first project using the technology at John Wheatley College in Glasgow (by ABK Architects).

The shatterproof nature of ETFE also makes it a viable choice under blast considerations because it eliminates the need to protect occupants from falling shattered glazing. Further, the ETFE assembly can be used as part of the larger overall blast mitigation strategy to effectively provide a "release valve" when blown out in the event of explosion.

ETFE is also noted to be durable, with an expected service life of between 20 to 25 years in architectural applications and some reports of the service life exceeding 30 years; nevertheless, regular maintenance by the manufacturer is needed to maintain this extended use. Following its serviceable life, the ETFE films can be recycled into products such as pipes, valves, or other components that do not have aesthetic or transparency requirements.

Case Studies

CASE STUDY 1

The first case study includes a domed ETFE skylight covering a large atrium space between two large office building structures constructed under a design-build contract for an institutional owner. The buildings are part of a larger campus in the Washington, DC, metropolitan area and have a combined footprint of over 2 million square feet. The owner hired an enclosure consultant at the beginning of the design, and the consultant remained involved in the project through substantial completion of the building enclosure. The ETFE pillows include three clear foils fritted for shading, and the skylight interfaces with polyvinyl chloride (PVC) roofing at several penthouses and a knee wall beneath the ETFE skylight (**figs. 5–8**).

Blast consideration was a primary driver for selecting ETFE for the skylight. A traditional large glass skylight could have been installed; however, the use of ETFE in the design eliminated the potential for falling glass should a blast event occur.

Because the ETFE was a delegated design, there were very few performance requirements in the project contract documents. However, there was a requirement for a full-scale laboratory mock-up of two "bays" of the ETFE roof. As required by the project manual, the laboratory testing procedure was submitted to the enclosure consultant for review. During review of the first submission, the consultant noted that the testing methodology and pass/fail criteria were the same as the typical methodology and criteria for glass-glazed aluminum-framed curtain walls. The consultant questioned the applicability of the tests and criteria, particularly the structural testing and associated deflection limitation requirements (L/240, the deflection limit criteria for glass). The deflection limitations appeared overly strict based on the consultant's understanding of the ETFE system. However, ETFE technology was still relatively new to construction in the United States and, at the time, the manufacturer had not developed a protocol to test performance criteria tailored to ETFE skylights.

FIG. 5 ETFE skylight interface with PVC at a penthouse.

FIG. 6 Conceptual section showing the ETFE skylight interface with PVC at a penthouse.

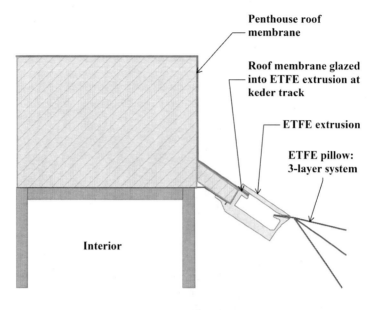

FIG. 7 ETFE skylight interface at the knee wall.

The submitted methodology and criteria were accepted by the manufacturer. Nevertheless, once representatives from the manufacturer, the design and construction team, the owner, and the enclosure consultant met at the testing facility, the representative from the manufacturer rejected the testing protocol, indicating that the foils could not meet the strict deflection criteria, necessitating an impromptu

FIG. 8 Conceptual section showing the ETFE skylight interface with PVC at the knee wall.

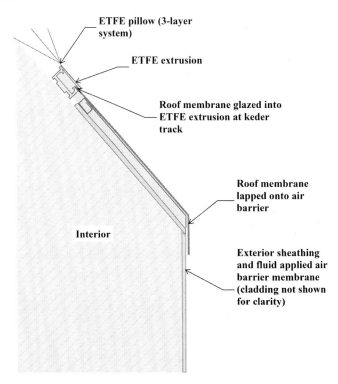

revision. The manufacturer confirmed that the pillows would deflect in excess of the submitted criteria as well as permanently deform under loads, although the deformation should decrease over time. The enclosure consultant recommended that iterative load cycles per ASTM E330, *Standard Test Method for Structural Performance of Exterior Windows, Doors, Skylights and Curtain Walls by Uniform Static Air Pressure Difference*, be performed to attempt to demonstrate the deflection tapering off, as indicated by the manufacturer (**fig. 9**). All parties agreed to this new methodology for the structural testing. A gauge was installed on the system interior (indicator location #10) and one on the system exterior (indicator location #9) (**fig. 10** and **table 2**). Deflections in the frame were minimal (≤ 0.13 in.).

The testing (**fig. 11**) confirmed the manufacturer's assertion that the permanent deflection of the exterior foil would decrease, and as the delta approached zero, the design team agreed that five load cycles were sufficient to show that the interior foil experienced limited permanent deflection and the exterior foil deflection was trending toward zero. After the standard load test a structural overload test was performed. At 117 psf positive load, a panel ruptured at a seam.

FIG. 9 ASTM E330 testing.

FIG. 10 Schematic of gauge locations for ASTM E330 testing.

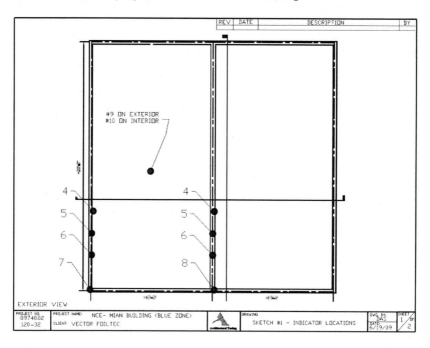

As the project moved into installation, the enclosure consultant developed a field testing and observation "checklist" protocol, focusing on the portions of the system that had the highest risk for failure. The list included extrusion splices

TABLE 2 Permanent set of panels after design load (five cycles) (permanent set in inches)

Indicator Location	Negative 62.0 psf
First load cycle	
9[a]	−3.04
10[b]	−0.05
Second load cycle	
9[a]	−1.24
10[b]	−0.37
Third load cycle	
9[a]	−0.68
10[b]	−0.03
Fourth load cycle	
9[a]	−0.44
10[b]	−0.03
Fifth load cycle	
9[a]	−0.25
10[b]	0.01

[a] A negative number indicates a movement of the panel toward the exterior.
[b] A negative number indicates a movement of the panel toward the interior.

FIG. 11 Graph of permanent deflection of ETFE recorded during ASTM E330 testing.

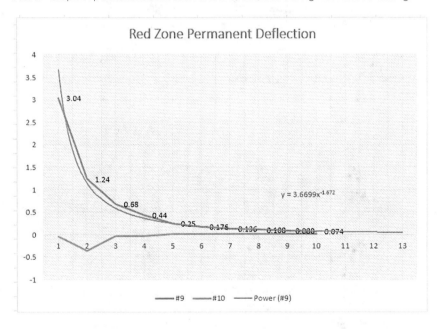

(**figs. 12–14**) due to the potential for air or water leakage if improperly sealed and relative inaccessibility once they are covered by cap extrusions as well as interface conditions.

During early installation, nozzle testing was performed at the extrusion splices in general accordance with AAMA 501.2, *Quality Assurance and Diagnostic Water Leakage Field Check of Installed Storefronts, Curtain Walls, and Sloped Glazing Systems* (**fig. 15**). The nozzle testing identified breaches in the silicone membrane bridge seal prior to final setting of the system. Peel adhesion testing of the silicone was also performed on select of the bridge seals (**fig. 14**). For this case study, all

FIG. 12 Extrusion splice.

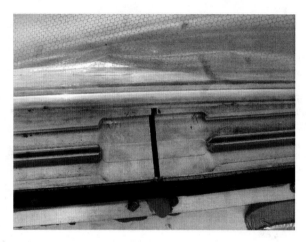

FIG. 13 Sealant for bridge seal.

FIG. 14 Bridge seal membrane.

FIG. 15 Nozzle testing of ETFE panel joints.

testing was performed at the base gutter extrusion due to limited access (per owner regulation, only employees from the manufacturer were allowed on top of the assembly). Testing of the assembled system (without the final cap) was also performed in accordance with AAMA 501.2 (by the manufacturer). After installation of the cap, the system was tested at interfaces using the spray rack described in ASTM E1105, *Standard Test Method for Field Determination of Water Penetration of Installed Exterior Windows, Skylights, Doors and Curtain Walls, by Uniform or Cyclic Static Air Pressure Difference* (**fig. 16**). The E1105 testing was performed at

FIG. 16 ASTM E1105 testing of ETFE panel joints (zero pressure differential).

zero pressure differential and with water sheeting over the system for 2 h (duration proposed by the construction team and agreed to by the design team and enclosure consultant). The rack was set in place by the manufacturer per the enclosure consultant's direction.

Nozzle testing did not identify any leaks; however, rack testing found leaks at interface conditions. All leaks identified occurred either at the PVC roofing membrane or at the interface between the PVC and ETFE. The design and construction team were ultimately able to understand this new assembly and develop a rational approach to quality assurance, which the enclosure consultant and the manufacturer applied to future projects, including the second case study. The manufacturer has also incorporated these quality-assurance measures into product literature.

CASE STUDY 2

The second case study reviews the quality-assurance procedures implemented at the ETFE assembly on a new train station in Anaheim, California. The facility contains approximately 200,000 square feet of ETFE pillows that serve as the primary enclosure system for the dome-shaped structure (**fig. 17**). The pillows are supported on large steel beams with steel chairs (**fig. 18**), which support the base extrusions and the "X" nodes of the ETFE system (**fig. 19**).

An enclosure consultant was engaged by the general contractor after issues arose during the mock-up phase, including tolerance issues with respect to the fit-up of the base extrusions that support the ETFE pillows and leaks that occurred during water testing of the mock-ups. The consultant's involvement began after the entire steel substructure, and approximately half of the base extrusions and pillows, had already been installed.

FIG. 17 Overall view of the support structure and ongoing installation of the ETFE pillows.

FIG. 18 Close-range view of the steel framing and chair supports for the ETFE base extrusions.

As part of the scope of services, the enclosure consultant performed an evaluation and review of the ongoing installation of the ETFE system, testing protocols, and field quality-control measures (being implemented by the manufacturer). The consultant's in-house difficult-access team was engaged to perform a limited visual review of accessible installed base extrusions and welds and continuity of the keder track and ETFE panel edges at close range via rope-access techniques. The limited

FIG. 19 Typical "X"-node base extrusion.

review scope also included peel adhesion testing at eight locations where the silicone bridge seals had been installed at the joints between the base extrusions (**fig. 20**), and isolated water testing using the methodology of AAMA 501.2 with modified equipment (due to water-access restrictions at the site) at the extrusion node joints. The consultant also performed limited water testing at the transition between the ETFE system and adjacent metal panels and the metal gutter at the base of the ETFE system.

FIG. 20 In-progress peel adhesion testing of the silicone bridge seal.

Overall, the limited visual assessment revealed isolated areas with damage to the base extrusions along the keder track, wrinkles, and splatter marks along the edges and within the field of the ETFE pillows, discontinuities between the individual pillows, and exposed locations of the pillow rope at discontinuous areas between the keder extrusions (**fig. 21**).

The adhesion testing of the silicone bridge seals indicated that they were well adhered to the substrate at all eight test locations, and no water intrusion was observed at the interior at the twelve sealant joints where water testing was performed.

Conclusions

In general, ETFE is a unique building enclosure solution that is durable, recyclable, transparent to UV and long-wave radiation, lightweight, and shatterproof. These attributes, when appropriately specified and designed for each project, application, and climate, have resulted in numerous successful installations. If not properly managed, however, the UV transparency that is optimal for plant growth can lead to fading of interior finishes, the solar heat gain coefficient (without coatings for frits) can result in overheated spaces, and the unattenuated noise can lead to a deafening interior during rain events. It is our understanding that the owners of the two case study facilities are satisfied with the performance of these unique assemblies as well as the service contracts provided by the manufacturer (which was initially of concern to the owners because it is a proprietary contract). Further, it is our understanding that the manufacturer in the case studies has incorporated the quality-control measures developed and successfully implemented on these two projects as part of their standard operating procedure and within their literature. Although not available at the time that the modified structural testing was developed in the first

FIG. 21 Visible portion of the pillow rope at a discontinuous keder extrusion.

case study, additional guidance for the design and construction for ETFE assemblies, including TensiNet guidelines for ETFE in Europe, is now available.

References

1. The Textile Institute, *Fabric Structures in Architecture* (Waltham, MA: Woodhead Publishing, 2015).
2. Vector Foiltec, "Project Lists," http://web.archive.org/web/20200221010312/https://www.vector-foiltec.com/projects-lists
3. L. A. Robinson, "Structural Opportunities of ETFE (Ethylene Tetra Fluoro Ethylene)" (master's thesis, Massachusetts Institute of Technology, 2005).
4. Vector Foiltec, "Texlon ETFE System," http://web.archive.org/web/20200220235854/https://www.vector-foiltec.com/texlon-etfe-system
5. A. Wilson, "ETFE Foil: A Guide to Design," http://web.archive.org/web/20200221000446/http://www.architen.com/articles/etfe-foil-a-guide-to-design
6. H. Poirazis, M. Kragh, and C. Hogg, "Energy Modelling of ETFE Membranes in Building Applications," in *IBPSA 2009* (Glasgow, UK: International Building Performance Simulation Association, 2009), pp. 696–703.
7. S. Bron-van der Jagt, C. Laudij, E. Gerretsen, E. Phaff, and T. Raijmakers, "Description of the Acoustic Characteristics of ETF Roof Structures," in *Proceedings of EuroNoise 2015* (Maastricht, the Netherlands: European Acoustic Association, 2015), pp. 1127–1132.

STP 1635, 2022 / available online at www.astm.org / doi: 10.1520/STP163520200123

Theresa A. Weston[1]

An Investigation of Stucco Wall Assembly Performance by Hygrothermal Simulation

Citation

T. A. Weston, "An Investigation of Stucco Wall Assembly Performance by Hygrothermal Simulation," in *Building Science and the Physics of Building Enclosure Performance: 2nd Volume*, ed. D. J. Lemieux and J. Keegan (West Conshohocken, PA: ASTM International, 2022), 71–89. http://doi.org/10.1520/STP163520200123[2]

ABSTRACT

Stucco walls are widely used, traditional building assemblies that have persisted by evolving to meet changing design criteria. In North America, stucco is a popular cladding choice both for its aesthetic appearance and for its inherent fire-resistant properties. Stucco use has increased in U.S. residential construction, comprising 27% of the U.S. primary exterior wall material in 2019, up from 16% 25 years earlier. Despite stucco's popularity, stucco systems have been associated with several performance questions, including cracking and water intrusion. These documented issues have spurred code and standard activity and product development over the last two decades. This study uses hygrothermal analysis to investigate several possible water intrusion mechanisms in stucco wall assemblies, including water absorption followed by inward vapor drive, water intrusion through stucco cracks, and bulk water intrusion at junctions or interfaces (e.g., window-wall interfaces). The analysis includes the relative severity of the moisture intrusion mechanism as a function of climate and assembly design. The assemblies considered were chosen considering recent and upcoming code and standard changes, including the addition of drainage and ventilation behind the stucco cladding. The efficacy of behind-cavity drainage and ventilation for mitigation of water intrusion mechanisms is analyzed.

Manuscript received December 21, 2020; accepted for publication May 3, 2021.
[1]The Holt Weston Consultancy, LLC, 2015 Westover Hills Blvd., Richmond, VA 23225, USA http://orcid.org/0000-0002-9822-7355
[2]ASTM Second Symposium on *Building Science and the Physics of Building Enclosure Performance* on April 24–25, 2022, and June 12, 2022 in Seattle, WA, USA.

Copyright © 2022 by ASTM International, 100 Barr Harbor Drive, PO Box C700, West Conshohocken, PA 19428-2959.

ASTM International is not responsible, as a body, for the statements and opinions expressed in this paper. ASTM International does not endorse any products represented in this paper.

Keywords

stucco, water-resistive barrier, rainscreen, moisture durability

Introduction

Stucco is a widely used and growing exterior cladding in residential construction, comprising 27% of the primary exterior wall material in 2019, up from 16% in 1995. Although used throughout the United States, stucco use is regional and is concentrated in the West (53% in 2019) and South (25% in 2019).[1] Despite stucco's popularity, issues with stucco applied over wood-based sheathing have been reported in industry reports in a variety of locations, including Vancouver, Canada; Woodbury, MN; Florida; and southeastern Pennsylvania.[2–7] In parallel to these reported issues has been an evolution in code and standard requirements for stucco applied over wood-based sheathing.[8] The International Code Council (ICC) code evolution continued through the most recent revision cycle, resulting in major changes in both the International Building Code (IBC) and International Residential Code (IRC). Climate zone-dependent requirements for enhanced drainage or ventilation behind the stucco cladding, or both, were included in the 2021 editions of both the IBC and IRC. The requirements in the 2021 IBC are as follows:[9]

> 2510.6 Water-resistive barriers. Water-resistive barriers shall be installed as required in Section 1403.2 and, where applied over wood-based sheathing, shall comply with Section 2510.6.1 or Section 2510.6.2.
>
> 2510.6.1 Dry climates. One of the following shall apply for dry (B) climate zones:
> 1. The water-resistive barrier shall be two layers of 10-minute Grade D paper or have a water resistance equal to or greater than two layers of water-resistive barrier complying with ASTM E2556, Type I. The individual layers shall be installed independently such that each layer provides a separate continuous plane and any flashing, installed in accordance with Section 1404.4 and intended to drain to the water-resistive barrier, is directed between the layers or a drainage space.
> 2. The water-resistive barrier shall be 60-minute Grade D paper or have a water resistance equal to or greater than one layer of water-resistive barrier complying with ASTM E2556, Type II. The water-resistive barrier shall be separated from the stucco by a layer of foam plastic insulating sheathing or other non-water-absorbing layer.
>
> 2510.6.2 Moist or marine climates. In moist (A) or marine (C) climate zones, water-resistive barrier shall comply with of one of the following:
> 1. In addition to complying with Item 1 or 2 of Section 2510.6.1, a space or drainage material not less than 3/16 inch (4.8 mm) in depth shall be applied to the exterior side of the water-resistive barrier.

2. In addition to complying with Item 2 of Section 2510.6.1, drainage on the exterior side of the water-resistive barrier shall have a minimum drainage efficiency of 90% as measured in accordance with ASTM E2273 or Annex A2 of ASTM E2925.

The 2021 IRC has parallel language.[10] It should be noted that either ventilation or enhanced drainage is required in moist and marine climates, whereas dry climates continue to require the former water-resistive barrier requirements between the stucco and the wood-based sheathing.

A review of the rationale statements for these code proposals in addition to the previously mentioned field reports reveals three major moisture transfer mechanisms in stucco assemblies, all of which may be reduced by the use of drainage and ventilation:

1. Inward vapor drive: This is a phenomenon in which water is absorbed by the stucco cladding and then is driven further into the wall assembly when the stucco is heated. This mechanism was investigated in detail by an ASHRAE-funded research project.[11]
2. Cracking: Stucco is expected to have some level of cracking, but when the cracking becomes excessive it will reduce the water resistance of the stucco plaster and place an increased water load on the stucco/water-resistive barrier interface. One historical industry reference states that cracking is not excessive if it is less than one lineal foot per 100 ft.^{2} of plaster and the width of cracks is not more than 50 mils wide. (For cracks more than the 50-mil width, the allowable lineal footage is reduced proportionately).[12] More recently, a review of crack acceptability criteria indicated that cracks 30 mils and less were "generally considered aesthetically acceptable" but also that "any visible cement plaster crack has the potential to allow bulk water intrusion."[13] A detailed study of the stucco mix, application, and curing found that poorly consolidated stucco had more shrinkage cracks and exhibited voids around the metal lath.[14] Water with dye was used to visualize the water migration pathway through the stucco. It was found that "the water uses the lath as a pathway disperse itself. That pathway is aided by voids around the lath wire that were created by the mix shrinking away from the lath." Shrinking and cracking of the mix was increased by using dirty sand or sand with high fine content, by neglecting to float the mix, or by neglecting to properly cure the stucco.
3. Bulk water intrusion at junctions and interfaces: This is when water bypasses the water-resistive barrier, for example, at the window-wall interface with insufficient flashing. Because this mechanism occurs because of the defects in the water management system, the best way to reduce it is improved quality control of the design and installation of these systems.

To understand the implications of the new drainage and ventilation requirements, hygrothermal analysis was conducted in a variety of climate zones. Simulations were designed to isolate the three aforementioned moisture transport mechanisms so that the interaction between the new code requirements and other code requirements could be examined and performance based on climate zone substantiated.

Hygrothermal Simulations

WUFI® 6.5 Pro was chosen as the simulation model because it is a well-validated and benchmarked model for hygrothermal applications. It is important to note that due to the inherent limitations of the model, results of the simulations were only expected to be predictive of relative performance and not specific assembly performance.

SIMULATION INPUTS

Exterior Climate

The wall assemblies were simulated for a 3-year time period for the following six locations/climate zones using weather files provided with the simulation software:
- Phoenix, AZ—climate zone 2B (desert/hot and dry)
- Los Angeles, CA—climate zone 3B (coastal/hot and dry)
- Tampa, FL—climate zone 2A (hot and humid)
- Philadelphia, PA—climate zone 4A (mixed)
- Seattle, WA—climate zone 4C (marine)
- Minneapolis, MN—climate zone 6A (cold)

Wall Assemblies

An example simulated assembly is shown in **figure 1**. Generic material data from the database included in the software program were used. The insulation thickness within the stud cavity (3.5 or 5.5 in.) was determined by prescriptive IECC-R-2018 requirements for the specific climate zone with one exception. Climate zone 6 (Minneapolis, MN) has a requirement for a continuous insulation in addition to the cavity insulation. The climate zone 6 case was modeled with 5.5 in. in stud cavity so that a direct comparison could be made with other cases and an additional model with 5.5-in. study cavity insulation with R-5 continuous foam plastic insulation. An interior vapor barrier/retarder was used in climate zones in which one is required, i.e., climate zones 4C (marine) and 6A (cold). However, because unintentional vapor retarders are sometimes included in assemblies even when they are not required, additional comparison simulations were conducted in which the assemblies included an interior vapor retarder. Air spaces were used both to introduce water (moisture source) and to incorporate ventilation into the assemblies (air change source).

Drainage and Ventilation

IBC-2021 allows two compliance options for meeting the enhanced drainage/ventilation requirements in moist and marine climate zones: "a space or drainage material not less than 3/16 inch (4.8 mm) in depth shall be applied to the exterior side of

FIG. 1 Simulated wall assembly design.

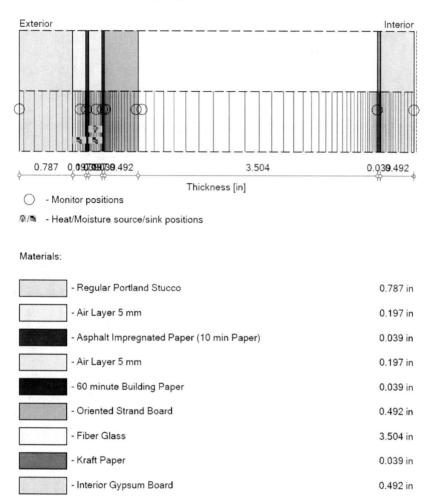

the water-resistive barrier" (option 1) or "drainage on the exterior side of the water-resistive barrier shall have a minimum drainage efficiency of 90% as measured in accordance with ASTM E2273 or Annex A2 of ASTM E2925" (option 2).

Ventilation rates for the simulation of these two options were based on reported research.[15,16] ASTM E2925, *Standard Specification for Manufactured Polymeric Drainage and Ventilation Materials Used to Provide a Rainscreen Function*,[17] provides a test method for evaluating air flow in rainscreen ventilation spaces. Measurements for using ASTM E2925 for rainscreen materials (described in option 1), enhanced drainage water-resistive barriers (described in option 2), and two layers of Grade D building paper (base case for dry climates) have been reported.[16]

Based on a review of these combined references, the air change rates used in the simulations to represent ventilation were as follows:
- Ventilation space, sometimes referred to a rainscreen (option 1): 100 air changes per hour (ACH)
- Textured water-resistive barriers (option 2): 5 ACH
- Two layers of building paper (traditional code requirement): 1 ACH

Moisture Intrusion

Each of the three moisture mechanisms studied used one of the following different water intrusion scenarios to simulate the specific mechanism:
- Inward vapor drive: Under these conditions, the only moisture source is water vapor diffusion.
- Cracking: Water based on wind-driven rain (wdr) was injected between the stucco and the most exterior water-resistive barrier layer. Levels of 1% wdr, which is the default level specified in ASHRAE Standard 160[18] (ASHRAE 160 default), and 3% wdr to represent a defective level of cracking were used in the simulations.
- Bulk water intrusion at interfaces: Water based on wdr was injected directly into the oriented strand board (OSB) sheathing to simulate bulk water bypassing the water-resistive barrier. This scenario simulates the occurrence of a defect in the wall's water management system. Levels of 1% wdr and 3% wdr were used in the simulations to represent two levels of bulk water intrusion severity. The direct injection of 1% and 3% wdr into the exterior sheathing has been used in previous studies to examine wall assembly performance in response to bulk water intrusion.[19,20] This assumption is used for modeling analysis and does not directly simulate real-world bulk water intrusion at interface defects.

Results and Discussion

Simulation results were analyzed on the basis of the OSB sheathing moisture content. A wide range in the moisture performance was found. The results are shown based on location (climate zone).

PHOENIX, AZ (CLIMATE ZONE 2B)

Results are shown in **figure 2**. None of the scenarios showed a year-over-year moisture accumulation, and moisture content remained less than 10% after the initial drying period. The results are consistent with not having a requirement beyond the long-standing code requirement for a performance equivalence of two layers of water-resistive barrier.

LOS ANGELES, CA (CLIMATE ZONE 3B)

Results are shown in **figure 3**. Although the OSB percentage moisture content was greater than that seen in the Phoenix, AZ simulations, none of the scenarios showed

FIG. 2 Phoenix, AZ simulation results.

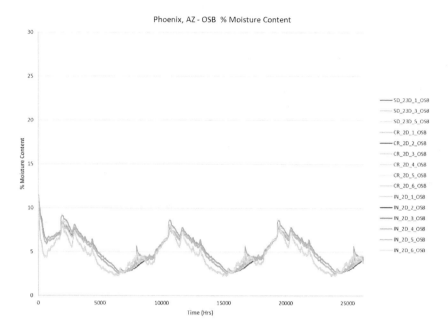

FIG. 3 Los Angeles, CA simulation results.

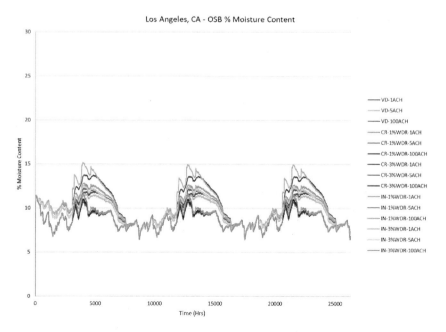

a year-over-year moisture accumulation, and the OSB moisture content remained less than 15%. The highest moisture content was observed when there was bulk water intrusion at the 3% wdr level, followed closely by the scenario in which there was infiltration at the 3% wdr level due to cracking. Ventilation at both 5 and 100 ACH reduced the moisture content, with 100 ACH virtually eliminating moisture gain by the OSB. Although performance was improved with the use of either a textured water-resistive barrier or a ventilation space, the overall low OSB moisture content across all of the simulations is consistent with not having a requirement beyond the long-standing code requirement for the performance equivalence of two layers of water-resistive barrier.

TAMPA, FL (CLIMATE ZONE 2A)

Results are shown in **figure 4**. Although none of the scenarios showed a year-over-year moisture accumulation, a high-risk moisture content reaching 25% was observed when there was bulk water intrusion at the 3% wdr level. OSB moisture content remained less than 15% for all of the remaining scenarios. As expected, inward vapor drive was a greater factor than in the dry climates. Annual moisture content peaks were greater than 10% but less than 15%. Scenarios with moisture entry due to cracking showed moisture contents between 10% and 15% and were significantly higher than inward drive alone. Ventilation reduced the moisture content at both 5 and

FIG. 4 Tampa, FL simulation results.

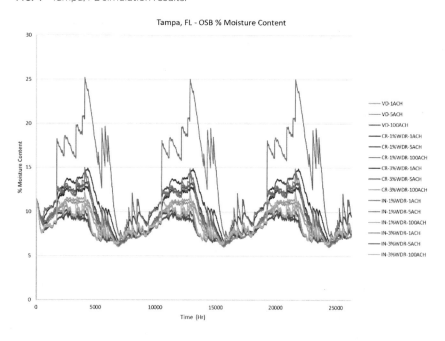

100 ACH, with the latter reducing moisture gain at a much greater level. The results show 2021 code requirements will help to mitigate water intrusion at defects but may be overly stringent when no bulk water intrusion is present.

PHILADELPHIA, PA (CLIMATE ZONE 4A)

Results are shown in **figure 5** (vapor drive scenarios), **figure 6** (cracking scenarios), and **figure 7** (bulk water intrusion scenarios). None of the scenarios showed a year-over-year moisture accumulation. Because climate zone 4A (which includes Philadelphia) is on the borderline of where interior vapor retarders are required by code, the inward vapor drive scenarios were run with and without an interior vapor retarder. Greater moisture content levels were evident when no vapor retarder was used than when an interior vapor retarder was present. The maximum moisture content (between 15% and 20% with no enhanced ventilation) occurred in the late winter. This indicates that moisture diffusing outward from the interior of the building is greater than inward drive. The moisture content was ameliorated with ventilation at 100 ACH but not at 5 ACH. Although interior vapor retarders are not required by code in this climate zone, it appears it would be an improved practice when ventilation is not provided.

The scenarios with moisture intrusion due to cracking show OSB moisture content that reached concerning levels: 20% at the 1% wdr level and 22% at the 3% wdr

FIG. 5 Philadelphia, PA vapor diffusion scenario simulation results.

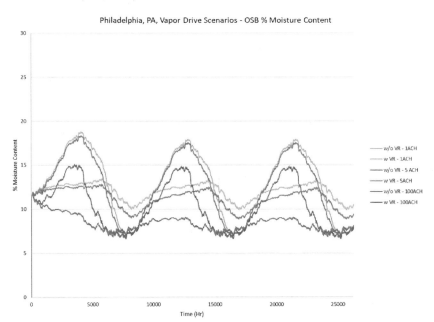

FIG. 6 Philadelphia, PA cracking scenario simulation results.

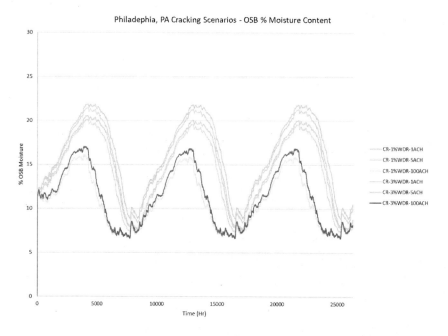

FIG. 7 Philadelphia, PA bulk water intrusion scenario simulation results.

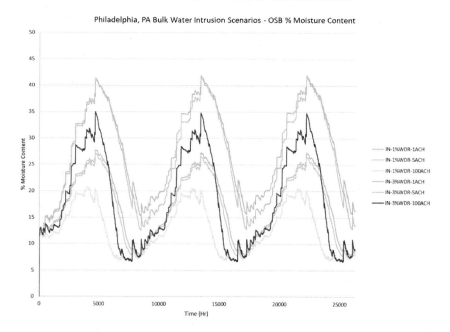

level. Ventilation at 100 ACH significantly reduced the OSB moisture content. The high OSB moisture content levels are because the moisture from intrusion at cracks adds to the background moisture from vapor diffusion from the interior of the building, as no internal vapor retarder was used in these simulations.

The scenarios with bulk water intrusion showed OSB moisture content at high-risk levels: greater than 25% for the 1% wdr level and peaks of greater than 40% for the 3% wdr level. Ventilation at 5 ACH showed only minimal effect. Ventilation at 100 ACH significantly reduced the OSB moisture content, but moisture content remained in a high-risk area, with OSB moisture content having peaks at 20% for the 1% wdr level and 35% for the 3% wdr level. As was observed for the cracking scenarios, the high OSB moisture content levels are likely due to the high background moisture from vapor diffusion from the interior of the building, as no internal vapor retarder was used in these simulations.

SEATTLE, WA

Results are shown in **figure 8** (vapor drive scenarios), **figure 9** (cracking scenarios), and **figure 10** (bulk water intrusion scenarios). The scenarios with only vapor drive showed no year-over-year OSB moisture accumulation. Peaks in moisture content just over 15% were observed. These peaks were effectively reduced with ventilation at either 5 or 100 ACH.

FIG. 8 Seattle, WA vapor drive scenario simulation results.

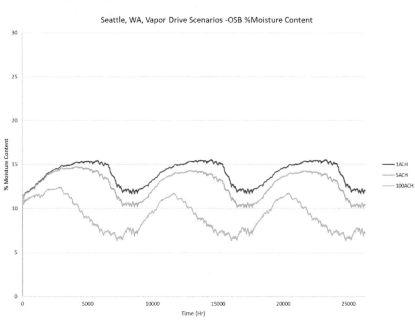

FIG. 9 Seattle, WA cracking scenario simulation results.

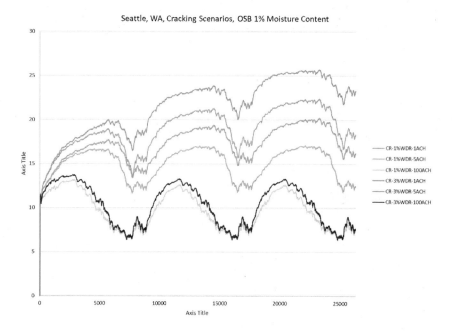

FIG. 10 Seattle, WA bulk water intrusion scenario simulation results.

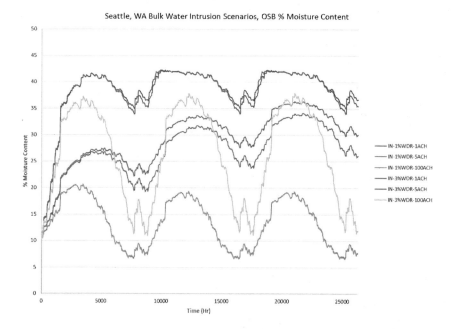

The scenarios with moisture intrusion due to cracking exhibited year-over-year moisture accumulation when no enhanced ventilation was provided. Moisture intrusion at cracking for the 1% wdr level and the 3% wdr levels produced OSB moisture peaks at 20% and 26%, respectively, with no enhanced ventilation. Ventilation at 5 ACH reduced the OSB moisture content. At the 1% wdr level, ventilation at 5 ACH eliminated the year-over-year accumulation and reduced the peak moisture content to 17%. At the 3% wdr level with 5 ACH, year-over-year moisture accumulation was still observed with peak moisture contents of 22%. Ventilation at 100 ACH significantly reduced the OSB moisture content, eliminating the year-over-year moisture accumulation and reducing the peak moisture content to below 15% for both the 1% wdr and 3% wdr intrusion levels.

The scenarios with bulk water intrusion showed year-over-year moisture accumulation and peak OSB moisture contents greater than 30% when either no enhanced ventilation or ventilation at 5 ACH was provided at both the 1% wdr and 3% wdr level. Ventilation at 100 ACH eliminated the year-over-year moisture accumulation at both the 1% and 3% wdr levels. The peak moisture content was reduced to 19% for 1% wdr but remained at a high-risk level (>35%) for the 3% wdr intrusion level.

MINNEAPOLIS, MN

Results are shown in **figure 11** (vapor drive scenarios), **figure 12** (cracking scenarios), and **figure 13** (bulk water intrusion scenarios). The scenarios with only vapor drive

FIG. 11 Minneapolis, MN vapor diffusion scenario simulation results.

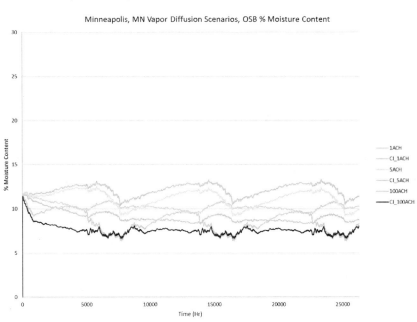

FIG. 12 Minneapolis, MN cracking scenario simulation results.

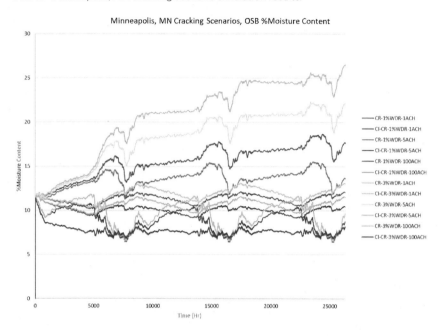

FIG. 13 Minneapolis, MN bulk water intrusion scenario simulation results.

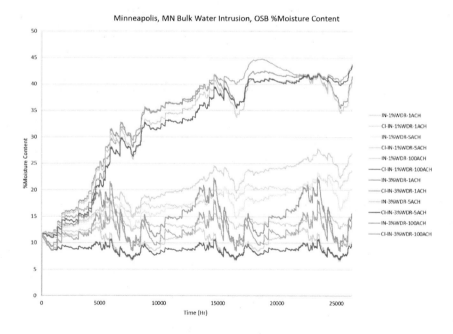

showed no year-over-year OSB moisture accumulation, and peaks in moisture content were all less than 15%.

The scenarios with moisture intrusion due to cracking exhibited year-over-year moisture accumulation when no enhanced ventilation was provided and when there was no exterior continuous insulation in the assembly. Moisture intrusion at cracking for the 1% and 3% wdr levels produced OSB moisture peaks of 19% and greater than 25%, respectively, with no enhanced ventilation. Ventilation at 5 ACH reduced the OSB moisture content. At the 1% wdr level, ventilation at 5 ACH eliminated the year-over-year accumulation and reduced the peak moisture content to 16%. At the 3% wdr level with ventilation at 5 ACH, year-over-year moisture accumulation was still observed, with peak moisture contents of 22%. Ventilation at 100 ACH significantly reduced the OSB moisture content, eliminating the year-over-year moisture accumulation and reducing the peak moisture content to below 15% for both the 1% and 3% wdr intrusion levels. When continuous insulation was added to the assemblies there was no year-over-year moisture accumulation, and the OSB peak moisture content was below 15% for all scenarios.

The scenarios with bulk water intrusion at the 3% wdr level showed year-over-year moisture accumulation and peak OSB moisture contents greater than 30% when either no enhanced ventilation or ventilation at 5 ACH was provided in assemblies both with and without exterior continuous insulation. Ventilation at 100 ACH eliminated the year-over-year moisture accumulation and reduced the peak moisture content below 25% for assemblies without exterior continuous insulation and below 20% for assemblies with exterior continuous insulation. The scenarios with bulk water intrusion at the 1% wdr level showed year-over-year moisture accumulation when either no enhanced ventilation or ventilation at 5 ACH was provided in assemblies both without exterior continuous insulation. In these cases, peak OSB moisture content was 27% and 23% in scenarios with no enhanced ventilation and ventilation at 5 ACH, respectively.

Ventilation at 100 ACH eliminated the year-over-year moisture accumulation and reduced the peak OSB moisture content below 15%. In the assemblies with exterior continuous insulation, year-over-year moisture accumulation with peak OSB moisture content was greater than 20% when there was no enhanced ventilation. In these assemblies, when ventilation at 5 ACH was provided, the year-over-year moisture accumulation leveled out, and the peak OSB moisture content was reduced to 15%. Ventilation at 100 ACH eliminated the year-over-year moisture accumulation and reduced the peak OSB moisture content below 15%.

RISK ANALYSIS OF CODE PROVISIONS

The simulation results were used to analyze the applicability of the new climate-based requirements. Scenarios were characterized from low to high risk as follows:
- Low risk: no year-over-year increase in OSB moisture content increase, and OSB moisture content remains below 15% MC; the 15% threshold is intentionally conservative because of the generic nature of the simulations

- Moderate risk: no year-over-year increase in OSB moisture content increase, and OSB moisture content has periods above 15% but stays below 20%
- High risk: year-over-year increase in OSB moisture content increase or OSB moisture content has periods above 20%, or both

Risk assessments are shown for "base moisture load" and "high moisture load" in **figure 14** and **figure 15**, respectively. The base moisture load risk assessment used the simulation results from the vapor drive scenarios and the cracking scenarios at 1% wdr. As previously mentioned, some cracking is inherent in stucco, and the 1% wdr is the default water deposition in ASHRAE 160.[18] Therefore, this level is considered a standard condition. The high moisture load risk assessment used the simulation results from the cracking scenarios at 3% wdr and the bulk water intrusion scenarios at 1% and 3% wdr. The cracking scenarios at 3% wdr represents a level of cracking that would be greater than expected under normal conditions. Bulk water intrusion at joints and interfaces would be dependent on the existence and extent of any defects in the water management system. Any assembly can be overwhelmed if

FIG. 14 Base moisture load risk assessment.

	Risk Assessment – Base Moisture Load					
	Vapor Drive			Cracking (1%wdr)		
	2-layers Traditional WRB	Drainage WRB	Ventilation Space	2-layers Traditional WRB	Drainage WRB	Ventilation Space
Phoenix (2B)	L	L	L	L	L	L
Los Angeles (3B)	L	L	L	L	L	L
Tampa (2A)	L	L	L	L	L	L
Philadelphia (4A)	M	M	L	M	M	M
Seattle (4C)	M	L	L	H	M	L
Minneapolis (6A)	L	L	L	H	M	L

FIG. 15 High moisture load risk assessment.

	Risk Assessment – High Moisture Load								
	Cracking (3%wdr)			Intrusion (1%wdr)			Intrusion (3%wdr)		
	2 layers Traditional WRB	Drainage WRB	Ventilation Space	2 layers Traditional WRB	Drainage WRB	Ventilation Space	2 layers Traditional WRB	Drainage WRB	Ventilation Space
Phoenix (2B)	L	L	L	L	L	L	L	L	L
Los Angeles (3B)	L	L	L	L	L	L	L	L	L
Tampa (2A)	L	L	L	L	L	L	H	L	L
Philadelphia (4A)	H	H	M	H	H	H	H	H	H
Seattle (4C)	H	H	L	H	H	M	H	H	H
Minneapolis (6A)	H	H	L	H	H	L	H	H	H

bulk water intrusion is large enough. Risk assessment is shown separately for water intrusion at 1% and 3% wdr.

Base Moisture Load Risk Analysis

In dry climates (Phoenix and Los Angeles), the traditional performance equivalence of the two-layer water-resistive barrier code requirement appears to be sufficient for good stucco performance. This supports the maintenance of traditional code requirements for dry climate zones. In climate zone 2A (Tampa), the code requirements were increased to require either a drainage water-resistive barrier or ventilation space. Based on the simulation results, raising the minimum code requirements in this climate zone appears not to be needed for the base moisture load levels. However, the simulations did not explicitly evaluate the value of increased drainage; they only evaluated the ventilation associated with products that provide increased drainage. Given the threat of extreme weather events in climate zone 2A, the additional drainage provided by these products may be warranted to provide added resilience. In climate zone 4A (Philadelphia), the vapor driven from the interior space seemed to dominate over inward vapor drive and may dominate over moisture intrusion at cracking. Added ventilation appears to have mitigated the vapor diffusion effects, but more work should be done to determine the combined effects of vapor retarder use and position with drainage and ventilation. In climate zones 4C (Seattle) and 6A (Minneapolis), the traditional performance equivalence of the two-layer water-resistive barrier requirement appears to be insufficient for acceptable wall assembly performance, and the new requirements appear warranted.

High Moisture Load Risk Analysis

It is important to consider that, by design, the high moisture load analysis represents moisture loads on the building assembly in excess of that covered by minimum code requirements. However, due to the high incidence of field failures reported for stucco construction, the new code requirements were examined to understand whether they would provide relief for higher levels of moisture intrusion. In dry climates (Phoenix and Los Angeles), the traditional performance equivalence of the two-layer water-resistive barrier code requirement appears to be sufficient to manage the threat from the moisture loads examined. Similarly, in climate zone 2A, the performance equivalence of the traditional two-layer water-resistive barrier code requirement appears to be sufficient to manage the threat from all but the highest levels of bulk water intrusion. However, as discussed above, the additional drainage provided by the new code requirements may be warranted due to threat of extreme weather events in this climate zone. In climate zones 4A, 4C, and 6A, high moisture loads are likely to overwhelm both the traditional performance equivalence of two-layer water-resistive barrier systems and enhanced drainage water-resistive barrier systems. Ventilation space systems provide some amelioration of high moisture loads but can be overcome by increased levels of bulk water intrusion.

Conclusions and Recommendations

The new code requirements for enhanced water-resistive barrier drainage and ventilation systems are predicted to provide improved stucco wall assembly performance and relief from limited water infiltration events. More specifically, the results of the simulation supported the following conclusions:

- A comparison of the moisture transfer mechanisms show the following severity ranking: bulk water intrusion > cracking > inward drive.
- The analysis confirms that stucco wall assembly performance is climate-specific, with lower risk in dryer and warmer climates.
- Ventilation increases performance against all moisture transfer mechanisms and is therefore considered a best practice. However, the level of added ventilation performance needs to be considered on a case-by-case basis to determine the best assembly considering other code provisions, local climate conditions, and affordability. For example, 5 ACH may be suitable in dry climates, whereas 100 ACH (or possibly higher) is necessary in marine and moist climates. More research is needed to understand the actual ventilation rates that are achieved with specific ventilation space and rainscreen assembly designs.
- Drainage and ventilation, although important to improving system performance, can be overwhelmed if defects allow water infiltration behind the water-resistive barrier. Proper water management detailing, especially at interfaces and joints, is essential to the successful performance of wall assemblies.
- Performance of stucco wall assemblies not only depends on the water-resistive barrier, drainage, and cladding ventilation design but is also affected by vapor retarder use and insulation selection. Further research to investigate design variants in stucco wall assemblies and in different climate zones is needed.
- As the need for greater building energy efficiency increases, wall assembly insulation strategies will evolve. These changes in insulation strategies will likely result in the changes to the requirements for other materials in the wall assemblies. Additional studies will need to be conducted as stucco wall assembly designs evolve.

References

1. U.S. Census Bureau, "Characteristics of New Housing," 2019, http://web.archive.org/web/20201218213407/https://www.census.gov/construction/chars
2. Morrison Hershfield Ltd., *Survey of Building Envelope Failures in the Coastal Climate of British Columbia* (Vancouver: Canada Mortgage and Housing Corporation, 1996).
3. D. Barrett, "The Renewal of Trust in Construction: Commission of Inquiry into the Quality of Condominium Construction in British Columbia," report submitted to the Government of British Columbia, Victoria, Canada, June 1998.

4. City of Woodbury Building Inspection Division, "Stucco in New Residential Construction—A Position Paper," 2011, http://web.archive.org/web/20201218215536/https://structuretech1.com/wp-content/uploads/2011/07/StuccoPositionPaper.pdf
5. M. Cramer, "Florida's Stucco Disaster: What Every Home Inspector Should Know Part 1," *ASHI Reporter*, May 2016, http://web.archive.org/web/20201218220232/https://www.homeinspector.org/Newsroom/Articles/Florida-s-Stucco-Disaster-What-Every-Home-Inspector-Should-Know-Part-1/14847/Article
6. M. Cramer, "Florida's Stucco Disaster: What Every Home Inspector Should Know Part 2," *ASHI Reporter*, June 2016, http://web.archive.org/web/20201218215743/https://www.homeinspector.org/Newsroom/Articles/Florida-s-Stucco-Disaster-Part-2/14877/Article
7. J. Lstiburek, "BSI-029: Stucco Woes—The Perfect Storm," *Building Science Insights*, January 22, 2010, http://web.archive.org/web/20201218220512/https://www.buildingscience.com/documents/insights/bsi-029-stucco-woes-the-perfect-storm
8. T. A. Weston, "Stucco Systems: A Review of Reported Data and Code and Standard Development" (paper presentation, 4th Residential Design & Construction Conference, State College, PA, February 2018).
9. International Code Council, Inc., *International Building Code* (Country Club Hills, IL: Author, 2021).
10. International Code Council, Inc., *International Residential Code* (Country Club Hills, IL: Author, 2021).
11. D. Derome, A. Karagiozis, and J. Carmeliet, "The Nature, Significance and Control of Solar-Driven Water Vapor Diffusion in Wall Systems—Synthesis of Research Project RP-1235" (paper presentation, ASHRAE 2010 Winter Meeting, Orlando, FL, January 2010).
12. J. J. Bucholtz, *The Consumer's Stucco Handbook* (San Jose, CA: Plaster Information Center, 1995).
13. J. Bowlsby, "Cement Plaster Metrics: Quantifying Stucco Shrinkage and Other Movements; Crack Acceptability Criteria for Evaluating Stucco" (paper presentation, Proceedings of the Building Envelope Technology Symposium, International Institute of Building Enclosure Consultants, San Antonio, TX, November 2010).
14. R. Webber, "The Stucco Solution: Creating Exterior Stucco Designed to Last a Lifetime," *Walls & Ceilings*, February 1997, 30–37.
15. E. Iffa and F. Tariku, "Application of a Cavity Ventilation Empirical Model for Hygrothermal Performance Assessment of Rain Screen Walls" (paper presentation, Buildings XIV Conference, Clearwater Beach, FL, December 2019).
16. T. Weston and M. Spinu, "Understanding the Benefit of Ventilation in Stucco Assemblies" (paper presentation, Buildings XIV Conference, Clearwater Beach, FL, December 2019).
17. *Standard Specification for Manufactured Polymeric Drainage and Ventilation Materials Used to Provide a Rainscreen Function*, ASTM E2925-19a (West Conshohocken, PA: ASTM International, approved August 1, 2019), https://doi.org/10.1520/E2925-19A.
18. *Criteria for Moisture-Control Design Analysis in Buildings*, ANSI/ASHRAE Standard 160-2016 (Atlanta, GA: ASHRAE, 2016).
19. T. A. Weston, "Using Hygrothermal Simulation to Assess Risk of Water Accumulation from Wall Assembly Defects," in *Building Walls Subject to Water Intrusion and Accumulation: Lessons from the Past and Recommendations for the Future*, ed. J. Erdly and P. Johnson (West Conshohocken, PA: ASTM International, 2014), 128–142, https://doi.org/10.1520/STP154920130058
20. A. Desjarlais and D. Johnston, "Energy and Moisture Impact of Exterior Insulation and Finish System Walls in the United States," in *Exterior Insulation and Finish Systems (EIFS): Performance, Progress and Innovation*, ed. P. Nelson and B. Egan (West Conshohocken, PA: ASTM International, 2016), 52–66, https://doi.org/10.1520/STP158520140097

STP 1635, 2022 / available online at www.astm.org / doi: 10.1520/STP163520210029

Zhe Xiao,[1,2] Michael A. Lacasse,[1] Maurice Defo,[1] and Elena Dragomirescu[2]

Moisture Load for a Vinyl-Clad Wall Assembly for Selected Canadian Cities

Citation

Z. Xiao, M. A. Lacasse, M. Defo, and E. Dragomirescu, "Moisture Load for a Vinyl-Clad Wall Assembly for Selected Canadian Cities," in *Building Science and the Physics of Building Enclosure Performance: 2nd Volume*, ed. D. J. Lemieux and J. Keegan (West Conshohocken, PA: ASTM International, 2022), 90–100. http://doi.org/10.1520/STP163520210029[3]

ABSTRACT

Moisture loads arising from the deposition of wind-driven rain (WDR) on building façades can bring about damage to wall assembly components and consequently affect their long-term performance. Hygrothermal simulations are often used by building practitioners to assess the risk to moisture damage within wall assemblies. Input of moisture loads to simulation models of wall assemblies use 1% of the WDR load deposited on the exterior surface of the wall as specified in ASHRAE Standard 160. However, results from several watertightness tests of wall assemblies have shown that the ratio between the WDR and moisture loads was highly dependent on the hourly WDR loads and driving-rain wind pressure (DRWP) acting on the wall surface. To permit correlating the varying WDR conditions with the moisture load for a vinyl-clad wall assembly, the wall was tested to a set of WDR intensities and DRWPs using a unique watertightness test apparatus, the Dynamic Wind and Wall Test Facility. The results from these tests

---

Manuscript received March 10, 2021; accepted for publication April 20, 2021.
[1]Construction Research Centre, National Research Council Canada, 1200 Montreal Rd., Ottawa, ON K1A 0R6, Canada Z X. https://orcid.org/0000-0001-8693-8515, M. A. L. https://orcid.org/0000-0001-7640-3701, M. D. https://orcid.org/0000-0001-9212-6599
[2]Dept. of Civil Engineering, 75 Laurier Ave. E, University of Ottawa, Ottawa, ON K1N 6N5, Canada E. D. https://orcid.org/0000-0001-8714-854X
[3]ASTM Second Symposium on *Building Science and the Physics of Building Enclosure Performance* on April 24–25, 2022, and June 12, 2022 in Seattle, WA, USA.

This work is not subject to copyright law. ASTM International, 100 Barr Harbor Drive, PO Box C700, West Conshohocken, PA 19428-2959.

ASTM International is not responsible, as a body, for the statements and opinions expressed in this paper. ASTM International does not endorse any products represented in this paper.

were thereafter used to formulate an empirical relation for the moisture load based on the hourly WDR load and DRWP as variables. The moisture load for this wall type was thereafter characterized on the basis of historical climate data in selected cities across Canada. It is supposed that the information derived from this study would offer both more accurate as well as more reliable information for input to simulation models.

Keywords
building facade, wall assembly, wind-driven rain, driving-rain wind pressure, watertightness test, moisture loads

Introduction

Moisture load is one of the factors that could induce detrimental effects to wall assemblies, such as frost damage,[1] salt migration,[2] and discoloration,[3] and is used when assessing the long-term performance of wall assemblies using hygrothermal simulation. Two approaches have currently been adopted that permit quantifying the moisture load in wall assemblies as implemented in the simulation model. A generic moisture load is considered 1% of the wind-driven rain (WDR) load deposited on the surface of the wall assembly, as given in ASHRAE Standard 160.[4] The other approach is to calculate the moisture load using water entry equations[5,6] derived from results of watertightness tests.[7] As stated in several studies,[8,9] the ratio (%) between the WDR load acting on wall assemblies and moisture load varies with the driving-rain wind pressure (DRWP) and WDR intensities that are applied to the exterior surface of the wall assembly, and these values for moisture load can differ significantly from 1% depending on the type of cladding and deficiency.

Existing water entry equations, equation (1)[5] and equation (2),[6] estimate the water entry rate based on WDR load and DRWP values. Both equations consider the water entry rate as a percentage of the WDR load. The percentages are determined by DRWPs and corresponding coefficients that are generated by fitting the results of watertightness test to DRWPs for different levels of the WDR load. In equation (1), the water entry rate has a linear relation to the WDR load, and the ratio between the moisture load and the WDR load is only affected by the DRWPs. However, according to the results of watertightness tests from which these water entry equations are derived, the water spray rate that simulates the WDR load could also affect the value of the ratio.

$$Q = m_p \times R_p \qquad (1)$$

where:
Q = water entry rate, L/min,
$m_p = A(\Delta P)^3 - B(\Delta P)^2 + C(\Delta P) + D$,
ΔP = pressure differences across the wall assembly, Pa,
A, B, C, and D = fitting factors, and
R_p = spray rate (equivalent to WDR intensity in L/min-m^2).

$$WE = WDR \times WE\% \qquad (2)$$

where:
 WE = water entry rate, L/h-m², and
 WDR = wind-driven rain intensity, L/h-m².

$$WE\% = a \times P^b \qquad (3)$$

where:
 $WE\%$ = percentage water entry for each cladding type,
 a and b = fitting factors, and
 P = pressure differences across the wall assembly, Pa.

In this study, a new, and innovative three-step approach is proposed to derive the ratio of the moisture load to the anticipated WDR load acting on the cladding of a wall assembly based on hourly WDR loads and average hourly DRWPs. This new approach requires fitting the results of watertightness tests to the values of both the water spray rates and DRWPs.

Accordingly, a 2.44 × 2.44-m vinyl-clad wall assembly was tested at the Dynamic Wind and Wall Testing Facility (DWTF) and subjected to a variety of water spray rates and DRWPs at National Research Council Canada (NRC). The watertightness test results were thereafter used to determine coefficients of the water entry equation from which the moisture load was calculated. Hourly WDR loads and DRWPs, taken between 1986 and 2016 for the cities of Vancouver, Calgary, Toronto, and Halifax, each of which represents four different climate zones, were implemented to the water entry equation to calculate the moisture load in a wall assembly for each of the respective locations being studied.

Methodology

TEST SPECIMEN

The wall assembly as tested was 2.44 × 2.44 m. A 600 × 600-mm window, a 100-mm-diameter circular ventilation duct, and an exterior electrical outlet box were installed to the exterior surface of the wall assembly, as shown in **figure 1**. In addition to the cladding, the wall assembly comprised a 30-minute asphalt-impregnated membrane, a 3/8-in. (10-mm) transparent polycarbonate sheathing representing the oriented strand board, 2 × 6 (51 × 152 mm) wood stud frames, and another 3/8-in. (10-mm) transparent polycarbonate sheathing representing the sheathing board and air barrier.

When the wall assembly was constructed in 2014, three 3-mm-diameter holes were bored at the top center of the caulking around the ventilation pipe, the bottom left of the caulking near the electrical outlet, and at the bottom corner of the window frame to simulate failures at these locations. Below these deficiencies, a yellow plastic mesh that covers the entire length of each appliance has been attached to the exterior surface of the water-resistance barrier to guide the water that has reached this layer to a metal water-collection trough underneath (**fig. 2A** and **B**) that can

FIG. 1 Tested vinyl-clad wall assembly.

FIG. 2 Water-collection troughs below the window (*A*); Water-collection troughs below the electrical outlet and ventilation duct (*B*).

(*A*)   (*B*)

further direct the collected water to reservoirs where the weight of the water could be measured. Prior to this test, those holes bored at different locations of the wall assembly were plugged, whereas all water-collection mechanisms in the wall assembly remained the same.

TEST FACILITY AND EXPERIMENTAL INPUT

The watertightness test was carried out using NRC's DWTF. The DWTF consists of (i) a pressure chamber connected to an industrial air blower from which positive pressure is applied to the exterior surface of the tested specimen, (ii) a rain-effect system that sprays water onto the test specimen to simulate WDR effects, and (iii) an exhaust flap system to allow adjustment of the differential pressure across the specimen.

Eight levels of dynamic pressure and four different water spray rates were implemented in the test. The magnitude of the dynamic pressure was determined on the basis of DRWP conditions in Canada, and the amplitude of the dynamic pressure was considered 20% of the magnitude of the pressure level. A frequency of 0.1 Hz was selected for varying the pressure according to the wind-velocity spectrum measured near the ground (<10 m).[10] Details of the experimental input are given in table 1.

CLIMATE DATA

The water entry equation derived from the results of the watertightness test was used to calculate the moisture load within the wall assembly based on the atmospheric WDR conditions anticipated on the exterior surface of a wall assembly. Hourly rainfall intensity and average hourly wind velocities, both of which are the primary parameters for deriving the WDR conditions, were collected for four Canadian cities (i.e., Calgary, Halifax, Toronto, and Vancouver) and then were used to assess the moisture load in a wall assembly situated in any of these four cities. The climate parameters were taken from measurements obtained at meteorological

TABLE 1 Experimental conditions

Steps	Dynamic Pressure Steps (Pa)	Spray Rates
1	$25 \pm 25 \times 0.2 \sin(2\pi ft)$	0.6 L; 1.3 L; 2.0 L; 2.7 L/min-m^2
2	$50 \pm 50 \times 0.2 \sin(2\pi ft)$	0.6 L; 1.3 L; 2.0 L; 2.7 L/min-m^2
3	$75 \pm 75 \times 0.2 \sin(2\pi ft)$	0.6 L; 1.3 L; 2.0 L; 2.7 L/min-m^2
4	$100 \pm 100 \times 0.2 \sin(2\pi ft)$	0.6 L; 1.3 L; 2.0 L; 2.7 L/min-m^2
5	$150 \pm 150 \times 0.2 \sin(2\pi ft)$	0.6 L; 1.3 L; 2.0 L; 2.7 L/min-m^2
6	$200 \pm 200 \times 0.2 \sin(2\pi ft)$	0.6 L; 1.3 L; 2.0 L; 2.7 L/min-m^2
7	$250 \pm 250 \times 0.2 \sin(2\pi ft)$	0.6 L; 1.3 L; 2.0 L; 2.7 L/min-m^2
8	$300 \pm 300 \times 0.2 \sin(2\pi ft)$	0.6 L; 1.3 L; 2.0 L; 2.7 L/min-m^2

Note: f = frequency (Hz); t: time (s).

stations operated by Environment and Climate Change Canada. The length of the weather data spans from 1986 to 2016.

Water Entry Equation

The proposed approach to obtain the water entry equation based on the results of a watertightness test includes three steps.

STEP 1: CORRELATE THE WATER ENTRY RATE TO THE WIND-DRIVEN RAIN PRESSURE INDEX

The water entry rate (mL/min) is calculated by dividing the amount of water (mL) that penetrates into the wall assembly during the watertightness test by the duration (min) of the test. The wind-driven rain pressure index (WDRPI) is a newly proposed indicator that permits describing the combined effects of the hourly WDR load and DRWP on the water entry rate as given in equation (4).

$$WDRPI = WDR^{\alpha} \times DRWP^{\beta} \qquad (4)$$

where:

$WDRPI$ = wind-driven rain pressure index,
WDR = hourly WDR load, L/h-m^2,
$DRWP$ = average hourly driving-rain wind pressure, Pa, and
α and β = adjustment coefficients.

The hourly WDR load in equation (4) is equivalent to the water spray rate used in the watertightness test, and the DRWP, calculated from knowledge of the wind velocity, is the pressure level applied to the specimen during the test. In terms of the watertightness test results, one water entry rate corresponds to one WDRPI, which is derived from the spray rate and applied pressure using equation (4). The adjustment factors α and β for the hourly WDR load and average hourly DRWP are generated by correlating the WDRPIs to the water entry rates.

STEP 2: OBTAIN PARAMETERS FOR WATER ENTRY EQUATION

An exponential function has been proposed to depict the relation between water entry rates and the WDRPIs as shown in equation (5). The parameters of the function are derived from fitting the plot of the water entry rates to the corresponding values for WDRPIs. The exponential function is thereafter used to calculate the water entry rates given the values of WDRPI.

$$\text{Water entry rate} = \alpha \times WDRPI^{\beta} \qquad (5)$$

where:

$WDRPI$ = wind-driven rain pressure index, and
α and β = adjustment coefficients.

The water entry results for the watertightness test, values of adjusted WDRPI, the water entry equation, and corresponding coefficients for the tested vinyl-clad wall assembly are shown in **figure 3**. The adjustment coefficient α, 1.544, for the

FIG. 3 Watertightness test results, adjusted WDRPI, and water entry equation.

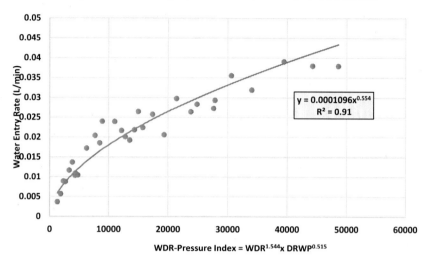

WDR load is larger than the adjustment coefficient β, 0.515, for the DRWP. This indicates that the water entry rate is more generally affected by the WDR load compared with the DRWP. The expressions as shown in equation (4) and equation (5) were selected from 14 different regression equations because they generated the least error and the highest correlation between the estimated water entry rates and measured water entry rates.

STEP 3: CALCULATE THE MOISTURE LOAD IN WALL ASSEMBLY USING CLIMATE DATA

Once the water entry equation is determined, the water entry rate of the wall assembly when exposed to the atmospheric WDR condition can then be calculated on the basis of the hourly values for the WDR load and DRWP. The ratio between the moisture load and hourly WDR load is then obtained by dividing the water entry rate by the hourly WDR load. The *t*-test results have shown that there is no significant difference between the water entry ratio measured from the watertightness test and the water entry ratio calculated using the proposed approach.

Moisture Load in Selected Cities

The hourly WDR loads in this study were calculated by equation (6), which is provided in ASHRAE Standard 160.[4]

$$\text{WDR load} = r_h \times F_E \times F_D \times F_L \times U \times \cos\theta \qquad (6)$$

where:
    WDR load = rain deposition rate on a vertical wall, L/m² · h,
    r_h = rainfall intensity on a horizontal surface, mm/h,
    F_E = rain exposure factor,
    F_D = rain deposition factor,
    F_L = empirical constant (0.2 kg · s),
    U = hourly average wind speed, m/s, and
    θ = angle between the wind direction and normal to the wall.

The rain exposure factor F_E and the rain deposition factor F_D were considered having, respectively, a value of 1 and 0.5. The values of other factors, except for the empirical constant F_L, were obtained from the climate data collected for the selected cities. The orientation of the wall assembly was determined on the basis of the prevailing wind direction during rain events for each city. In addition, as shown in equation (7), the DRWPs were calculated from the wind velocity during rain events using Bernoulli's principle assuming that the air density was 1.225 kg/m³.

$$DRWP = \frac{1}{2}\rho v^2 \qquad (7)$$

where:
    $DRWP$ = driving – rain wind pressure, Pa,
    ρ = air flow density (1.225 kg/m³), and
    v = average hourly wind velocity during rain events, m/s.

The moisture loads for the tested vinyl-clad wall assembly when it was situated in each of the selected cities is shown in **figure 4**. Each subplot includes a histogram for different levels of moisture load derived from each rain event, as well as the cumulative distribution of the histogram. The white bar plot represents the moisture load estimated using the proposed water entry equation, whereas the shaded bar plot represents the moisture load considered 1% of the WDR load.

For all cities that were analyzed, the frequencies of the moisture load equivalent to 1% of the WDR load, and that occur within a range of 0 and 0.002 L/m²-h, were much higher than the frequencies of moisture load estimated by the water entry equation. With the exception of the moisture load derived for Halifax, and that ranged in value between 0.002 and 0.004 L/m²-h, the frequencies of the moisture load estimated by the water entry equation were slightly higher than that for a moisture load equivalent to 1% of the WDR load. These observations suggested that the water entry equation generated a higher value for the moisture load in the wall assembly under the same WDR condition. The cumulative distributions for each category of moisture load also indicated that the water entry equation generated larger values of moisture load compared with that using 1% of the WDR load at the same percentile.

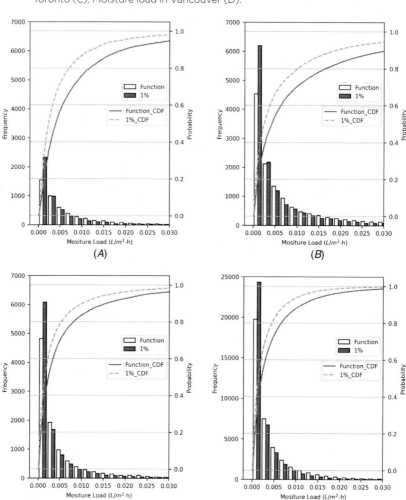

FIG. 4 Moisture load in Calgary (*A*); Moisture load in Halifax (*B*); Moisture load in Toronto (*C*); Moisture load in Vancouver (*D*).

The magnitudes of moisture load at different percentiles in each city are given in **table 2**. For all categories analyzed and cities studied, moisture loads derived from the proposed water entry equation were greater than the moisture load equivalent to 1% of the WDR load. Although Calgary had the fewest rain events during the period analyzed (i.e., 1986–2016), as shown in **figure 4**, the magnitude of moisture load in the wall assembly was higher than that in other cities under the same percentile value. Moisture loads for the wall assembly in Vancouver were lower than those in other cities, although rain events in Vancouver were those that occurred most frequently. The tested wall assembly if located in Halifax would

TABLE 2 Moisture load for different percentiles

Categories	Moisture Load	Calgary	Halifax	Toronto	Vancouver
Mean	1% of WDR	0.0052[a]	0.0039	0.0023	0.0009
	Function	0.0085	0.0064	0.0039	0.0015
	Dif. (%)	163%	164%	170%	167%
Percentile					
50	1% of WDR	0.0023	0.0014	0.0008	0.0004
	Function	0.0039	0.0025	0.0015	0.0008
90	1% of WDR	0.0122	0.0098	0.0056	0.0019
	Function	0.0196	0.0159	0.0093	0.0032
95	1% of WDR	0.019	0.0156	0.0093	0.0029
	Function	0.0312	0.0252	0.0149	0.0049
99	1% of WDR	0.0424	0.0345	0.0238	0.0066
	Function	0.0664	0.0548	0.0366	0.0117
99.5	1% of WDR	0.0551	0.0493	0.0312	0.0085
	Function	0.0901	0.0714	0.047	0.0145

[a] unit = L/m^2-h.

expect more frequent rain events and a higher moisture load than that of Toronto. The water entry equation indicated average ratios of 1.63% to 1.7% between the moisture in the tested vinyl-clad wall assembly and the WDR load that deposited on its exterior surface.

Conclusions

In this paper we proposed a three-step approach that permits calculating the moisture load for any tested wall assembly given the hourly WDR load and average hourly DRWP, both of which are available from meteorological stations. A water entry equation was established based on the results of a watertightness test for a vinyl-clad wall assembly. The equation was used to calculate the moisture load for the tested wall assembly when it was situated in selected Canadian cities. It was concluded that the moisture load derived from the water entry equation was, on average, 63% to 70% greater than a moisture load equivalent to 1% of the WDR for all cities studied in this paper.

This three-step approach can also be used to estimate the moisture load in other types of wall assemblies. To determine the adjustment coefficients of the water entry equations, additional watertightness tests need to be implemented to wall assemblies of interest because the results of watertightness tests are affected by the configurations of the wall assembly. Deficiencies for each type of wall assembly also need to be standardized to permit comparison of moisture loads in different types of wall assemblies.

ACKNOWLEDGMENTS

This work is part of the Climate-Resilient Buildings and Core Public Infrastructure project supported by National Research Council Canada's Construction Research Centre and Infrastructure Canada.

References

1. K. Van Balen, "Expert System for Evaluation of Deterioration of Ancient Brick Masonry Structures," *Science of the Total Environment* 189–190 (1996): 247–254, https://doi.org/10.1016/0048-9697(96)05215-1
2. A. E. Charola and L. Lazzarini, "Deterioration of Brick Masonry Caused by Acid Rain," in *Materials Degradation Caused by Acid Rain*, ed. R. Baboian (Washington, DC: American Chemical Society, 1986), 250–258, https://doi.org/10.1021/bk-1986-0318.ch017
3. L. Franke, I. Schumann, R. Hees, L. V. D. Klugt, S. Naldini, K. Balen, and J. M. Mateus, *Damage Atlas, Classification and Analysis of Damage Patterns Found in Brick Masonry* (Stuttgart, Germany: Fraunhofer IRB Verlag, 1995).
4. *Criteria for Moisture Control Design Analysis in Buildings*, ASHRAE Standard 160P (Atlanta, GA: ASHRAE, 2016).
5. N. Sahal and M. A. Lacasse, "Water Entry Function of a Hardboard Siding-Clad Wood Stud Wall," *Building and Environment* 40, no. 11 (2005): 1479–1491, https://doi.org/10.1016/j.buildenv.2004.11.019
6. T. V. Moore, M. A. Lacasse, and M. Defo, "Determining Moisture Source Due to Wind Driven Rain for Input to Hygrothermal Simulations Using Experimental Methods" (paper presentation, 2019 ASHRAE Annual Conference, Kansas City, MO, June 22–26, 2019).
7. M. A. Lacasse, T. O'Connor, S. C. Nunes, and P. Beaulieu, "Report from Task 6 of MEWS Project: Experimental Assessment of Water Penetration and Entry into Wood-Frame Wall Specimens - Final Report," Research Report No. RR-133, National Research Council Canada Institute for Research in Construction (2003), https://doi.org/10.4224/20386351
8. M. A. Lacasse, N. Van Den Bossche, S. Van Linden, and T. V. Moore, "A Brief Compendium of Water Entry Results Derived from Laboratory Tests of Various Types of Wall Assemblies," *MATEC Web Conference* 282 (2019): 02050, https://doi.org/10.1051/matecconf/201928202050
9. N. Van Den Bossche, M. Lacasse, and A. Janssens, "Watertightness of Masonry Walls: An Overview" (paper presentation, 12th International Conference on Durability of Building Materials and Components, Porto, Portugal, April 12, 2011).
10. H. W. Teunissen, *Characteristics of the Mean Wind and Turbulence in the Planetary Boundary Layer* (Toronto, Canada: Institute for Aerospace Studies, University of Toronto, 1970).

Derek J. Ziese[1]

Design and Detailing Air and Weather Barrier Transitions to Fenestrations

Citation

D. J. Ziese, "Design and Detailing Air and Weather Barrier Transitions to Fenestrations," in *Building Science and the Physics of Building Enclosure Performance: 2nd Volume*, ed. D. J. Lemieux and J. Keegan (West Conshohocken, PA: ASTM International, 2022), 101–132. http://doi.org/10.1520/STP163520210034[2]

ABSTRACT

Detailing the 21st-century high-performance building has grown more complicated. Energy-efficiency goals have led to increased continuous insulation thickness, rainscreen claddings, and accommodations for cavity ventilation and drainage. These changes have resulted in deeper wall systems and increased importance on the air and weather barrier and wall flashings as a weatherproofing system. Code requirements for the exterior wall assembly, particularly those outlined in NFPA 285, impact the types of insulation, air barriers, and flashings/closures configurations not requiring custom-tested configurations. These challenges have brought about significant changes to detailing at fenestrations. Meanwhile, the glazing industry has not provided a uniform or standard approach to air barrier transitions to opaque wall systems, given the numerous fenestration types and configurations available. Due to the complicated nature of detailing the myriad of systems and configurations available, designers often denote generic transition details at the critical junction of the wall and fenestration systems. Designs of the fenestration transition are often delayed to the submittal stage when the final system selections are made often after value engineering exercises. If the proposed fenestration system and building exterior design intent are not adequately evaluated, poor performance issues may result in air and water infiltration. Depending upon the complexity of the design,

---

Manuscript received March 22, 2021; accepted for publication June 20, 2021.
[1]Building Enclosure and Consulting, Gale Associates, Inc., 122 Kenilworth Dr., Ste. 206, Towson, MD 21204, USA
[2]ASTM Second Symposium on *Building Science and the Physics of Building Enclosure Performance* on April 24–25, 2022, and June 12, 2022 in Seattle, WA, USA.

Copyright © 2022 by ASTM International, 100 Barr Harbor Drive, PO Box C700, West Conshohocken, PA 19428-2959.

ASTM International is not responsible, as a body, for the statements and opinions expressed in this paper. ASTM International does not endorse any products represented in this paper.

mock-ups are recommended to allow modifications of flashings and seals prior to first installations on the building. The overall relationship of the cladding, insulation, air barrier, flashings, and fenestration must be considered in the design phase because compatibility, functionality, and constructability of the transitions may dictate different fenestration configurations or air barrier transition materials, or both.

Keywords
detailing, fenestration, air barrier, weather barrier, rainscreens

Introduction

The primary function of the building enclosure is to protect the interior occupants from the exterior environment. With the use of mechanical conditioning systems, the term "protection" has evolved from defense against weather to include segregation of the interior and exterior environments and full control of the interior environment. This includes the control of moisture migration, air leakage, and thermal differences between the interior and exterior for which the exterior wall plays a vital role. Two critical requirements of the exterior wall are to stop uncontrolled air flow and bulk water migration. To accomplish this, control layers such as air and weather barriers (AWBs), drainage planes, and associated flashing components are used. To be functional, these systems must be continuous across the wall assembly and connect to penetrations such as fenestrations.

Fenestrations are openings in an opaque wall and typically consist of glazing assemblies intended to provide natural light and ventilation to the interior. There are a multitude of glazing types used in today's construction such as curtain wall, window, and storefront assemblies. Each assembly has unique performance requirements and fabrication configurations, but most exterior systems are designed to limit air infiltration and manage moisture. To provide continuous control layers through a fenestration, the transition assembly between the fenestration and wall must be carefully detailed. Proper detailing and integration of specific fenestration assemblies into the wall requires designers to juggle a multitude of specific requirements for both the opaque wall and the fenestration. Specifically, designers must consider how their detailing will impact the performance dynamics of the wall and make a determination of the constructability during the design phases, often without knowledge of the actual methodologies and sequencing to be used in construction.

I have been engaged on numerous projects in which control layer transitions to fenestration are not fully developed during the design, resulting in detail creation in the construction phase where there is little time allocated to the issue and can have significant schedule and cost demands. This paper intends to provide the design professional with an understanding of the requirements of the air and bulk moisture control layers and their coordination with commercial fenestration systems. Discussions include various opaque wall assemblies, relevant code requirements, and general design principles to maintain continuity of the control layers through typical fenestration types.

Opaque Wall Materials and Assemblies

CONTROL LAYERS

The building enclosure must be designed to control various elements to include bulk moisture, air leakage, vapor transmission, and thermal differences.[1] As such, control, or "separation," layers are used in opaque wall assemblies to manage these elements and related forces. Control layers in an exterior wall assembly include the following:
- Water-resistive barrier (WRB)—protection against bulk rainwater and typically performs as the drainage plane
- Air barrier—control air leakage through the wall assembly
- Thermal barrier (insulation layer)—minimize heat transfer and potential for condensation
- Vapor retarder—control of water vapor diffusion

Singular components within the exterior enclosure assembly may serve as one or more of the control layers, depending on its properties and location. For the purposes of this paper, the air barrier described is also the WRB, or drainage plane, and is located on the exterior side of the stud wall or mass backup wall.

When specifying the AWB, multiple membrane types and material properties are available depending on desired application and substrate type. These membranes are primarily divided into the following types:
- Self-adhered sheet membranes are available as vapor-impermeable or vapor-permeable air and water barriers.
- Fluid-applied membranes are available as vapor-impermeable or vapor-permeable air and water barriers. These can be roller- or spray-applied. Recent advances have made sheathing with preapplied AWB membranes available; the sheathing joints are sealed, typically with an appropriate transition membrane.
- Building wraps can serve as vapor-permeable air and water barriers but must be mechanically attached to resist building design pressures.
- Spray urethane foams (SPUFs) can be utilized as continuous insulation and as an air and water barrier membrane. Depending on installation thickness, SPUFs can be considered a vapor retarder.
- Rigid insulation with taped joints can be utilized as continuous insulation and as an air and vapor barrier membrane.

For most systems listed above, the AWB membrane is turned into the rough opening via stripping membranes. These are typically in the form of self-adhered modified bitumen sheets with polyethylene facers and are used in combination with termination mastics and sealants, or liquid flashing membranes. Because most commercial fenestrations utilize silicone sealants, compatibility of the above should be confirmed with the fenestration manufacturer.

WALL ASSEMBLIES

The complexity of wall assemblies has changed dramatically from historic monolithic walls where the masonry served as the cladding, water reservoir, and thermal mass to modern walls that are designed with multiple components such as rainscreen cladding systems, air barriers, and continuous insulation. The hygrothermal behavior of the wall will dictate where the control layers should be located and therefore how and where each control layer connects to the fenestration. Opaque exterior wall assemblies can be classified as reservoir, barrier, and rainscreen wall systems. New commercial buildings typically utilize barrier or rainscreen wall assemblies over a metal-framed or mass backup wall.

Barrier Wall Assemblies

Barrier walls are designed as fully sealed systems that prevent the migration of bulk moisture and air infiltration at the exterior face of the building. These systems typically rely on an impervious cladding or finish material with sealant at the assembly joints and are often face-sealed systems where the outer layer serves as the air and moisture barrier. Face-sealed systems include certain types of exterior insulated finish systems, some metal wall panels, and some precast concrete wall panel systems. Barrier walls can include joints between panels with dual-stage sealant joints. The outer seal can be weeped, but the inner seal must be continuous for the barrier wall concept to be realized. Gasketed barrier walls are also common and can consist of interlocking modular units such as prefabricated panelized walls or insulated metal panels. Because the cladding and AWB are one assembly in barrier walls, the systems can involve less components and trade contractors; however, they are reliant on a singular seal for moisture penetration with little redundancy and are therefore highly workmanship-dependent. Deficiencies in the face seals could result in moisture infiltration.

Rainscreens

The term "rainscreen" refers to an open cladding that permits the transport of bulk moisture to the subsequent layers of the wall system. A typical rainscreen will include a cladding system in front of a WRB resulting in a drainable cavity between the two. Rainscreens are also referred to as cavity walls or drained wall systems. Due to the open nature of the rainscreen systems, the rainscreen is heavily reliant on the moisture barrier and flashing systems to control moisture paths and to direct moisture out of the system.

Rainscreens can be further divided into drained and back-ventilated (DBV) or pressure-equalized (PE) assemblies. The DBV assemblies are designed to allow moisture past the cladding systems and evacuate moisture from the cavity via flashings; evaporation of the cavity is achieved through back ventilation. PE assemblies are designed with compartmentalization of the drainage cavity to achieve pressure equalization and therefore the cladding joints theoretically are open to air but not moisture. PE assemblies rely on a continuous air barrier for equalization of the compartments. Design and construction deficiencies often result in PE systems that are not fully equalized and allow moisture ingress through the cladding to the cavity. Therefore, similar detailing to that of the DBV systems regarding flashings and moisture barriers are used.

CODE AND PERFORMANCE REQUIREMENTS

As with most building design processes, the code and owner requirements typically dictate the baseline for detailing. Unfortunately, most codes are not prescriptive on how to handle transitions at fenestrations other than requiring that AWBs be continuous across a wall assembly and that flashings shall be provided at fenestrations perimeters. The primary applicable national codes and references that relate to the AWB are the International Building Code (IBC)[2] Chapter 14, "Exterior Walls, International Energy"; Conservation Code (IECC)[3] Chapter 4, "Commercial Energy Efficiency"; and National Fire Protection Association (NFPA) 285, *Standard Fire Test Method for Evaluation of Fire Propagation Characteristics of Exterior Wall Assemblies Containing Combustible Materials.*[4] Refer to **tables 1** and **2** for excerpts from the codes that relate specifically to the AWB to fenestration transition.

The IECC states in Chapter 4 that the maximum air leakage of the individual materials, wall assemblies, and whole building shall be 0.004, 0.04, and 0.4 cfm/sq. ft., respectively. These values are stated in Sections C402.5, C402.5.1.2.1, and C402.5.1.2.2, respectively. There are three compliance methods (see **table 3**) with two prescriptive compliance methods that use manufacturer-tested materials or assemblies and a third compliance path that requires whole building air-leakage testing.

In addition to the opaque wall air-leakage rates mentioned above, fixed fenestrations have a maximum allowable air-leakage rate of 0.06 cfm/sq. ft. Depending on the type of system, the maximum air-leakage rate from the manufacturer may be tested at 1.57 lb./sq. ft. (kg/m^2) or 6.24 lb./sq. ft. (kg/m^2), when tested in accordance with ASTM E283, *Standard Test Method for Determining Rate of Air Leakage through Exterior Windows, Skylights, Curtain Walls, and Doors under Specified Pressure Differences across the Specimen.*[9] Field quality-control testing should be in accordance with ASTM E783, *Standard Test Method for Field Measurement of Air Leakage through Installed Exterior Windows and Doors.*[10] The difference in air-leakage rates between the AWB and the fenestration can create confusion for the designer as to what the required performance value of the transition assembly should be. The transition assembly should be designed to the more stringent of the fenestration and AWB air-leakage requirements. Regarding water penetration, the fenestration will typically have published resistance values when tested in accordance with ASTM E331, *Standard Test Method for Water Penetration of Exterior Windows, Skylights, Doors, and Curtain Walls by Uniform Static Air Pressure Difference*,[11] or ASTM E547, *Standard Test Method for Water Penetration of Exterior Windows, Skylights, Doors, and Curtain Walls by Cyclic Static Air Pressure Difference*,[12] if the system is common. Some systems may also have dynamic testing data per AAMA 501.1, *Water Penetration of Windows, Curtain Walls and Doors Using*

TABLE 1 Summary of selected IBC AWB requirements

Section	Title	Requirement
1402.2	Weather protection	Exterior walls shall provide the building with a weather-resistant exterior wall envelope. The exterior wall envelope shall include flashing as described in Section 1404.4. The exterior wall envelope shall be designed and constructed in such a manner as to prevent the accumulation of water within the assembly by providing a water-resistive barrier behind the exterior veneer, as described in Section 1403.2, and a means of draining water that enters the assembly to the exterior. Protection against condensation in the exterior wall assembly shall be provided in accordance with Section 1404.3.
1402.3	Structural	Exterior walls, and the associated openings, shall be designed and constructed to resist safely the superimposed loads required by Chapter 16.
1402.4	Fire resistance	Exterior walls shall be fire-resistance rated as required by other sections of this code, with opening protection as required by Chapter 7.
1402.5	Vertical and lateral flame propagation	Exterior walls on buildings of Type I, II, III, or IV construction that are greater than 40 ft. (12,192 mm) in height above grade plane and contain a combustible water-resistive barrier shall be tested in accordance with and comply with the acceptance criteria of NFPA 285. For the purposes of this section, fenestration products, flashing of fenestration products, and water-resistive barrier flashing and accessories at other locations, including through-wall flashings, shall not be considered part of the water-resistive barrier.
1403.2	Water-resistive barrier	Not fewer than one layer of No. 15 asphalt felt, complying with ASTM D226, *Standard Specification for Asphalt-Saturated Organic Felt Used in Roofing and Waterproofing*, for Type I felt or other approved materials, shall be attached to the studs or sheathing, with flashing as described in Section 1404.4, in such a manner as to provide a continuous water-resistive barrier behind the exterior wall veneer.
1404.4	Flashing	Flashing shall be installed in such a manner so as to prevent moisture from entering the wall or to redirect that moisture to the exterior. Flashing shall be installed at the perimeters of exterior door and window assemblies, penetrations and terminations of exterior wall assemblies, exterior wall intersections with roofs, chimneys, porches, decks, balconies and similar projections, and at built-in gutters and similar locations where moisture could enter the wall. Flashing with projecting flanges shall be installed on both sides and the ends of copings, under sills, and continuously above projecting trim. Where self-adhered membranes are used as flashings of fenestration in wall assemblies, those self-adhered flashings shall comply with AAMA 711. Where fluid applied membranes are used as flashing for exterior wall openings, those fluid applied membrane flashings shall comply with AAMA 714.
1404.4.1	Exterior wall pockets	In exterior walls of buildings or structures, wall pockets or crevices in which moisture can accumulate shall be avoided or protected with caps or drips, or other approved means shall be provided to prevent water damage.

TABLE 2 Summary of selected IECC AWB requirements

Section	Title	Requirement
C402.5.1	Air barriers	A continuous air barrier shall be provided throughout the building thermal envelope. The air barriers shall be permitted to be located on the inside or outside of the building envelope, located within the assemblies composing the envelope, or any combination thereof. The air barrier shall comply with Sections C402.5.1.1 and C402.5.1.2. An exception is provided for Climate Zone 2B, where air barriers are not required.
C402.5.1.1	Air barrier construction	The continuous air barrier shall be constructed to comply with the following: 1. The air barrier shall be continuous for all assemblies that are the thermal envelope of the building and across the joints and assemblies. 2. Air barrier joints and seams shall be sealed, including sealing transitions in places and changes in materials. The joints and seals shall be securely installed in or on the joint for its entire length so as not to dislodge, loosen, or otherwise impair its ability to resist positive and negative pressure from wind, stack effect, and mechanical ventilation.

TABLE 3 Compliance methods

Compliance Option	Test Method	Summary	Failure Criteria
1	ASTM E2178, *Standard Test Method for Determining Air Leakage Rate and Calculation of Air Permeance of Building Materials*[5]	Material lab test: tests only singular AB materials	0.004 cfm/sq. ft. at 75 Pa (0.3" water column)
2	ASTM E2357, *Standard Test Method for Determining Air Leakage Rate of Air Barrier Assemblies*[6]	Assembly lab test or field mock-up: measures air permeance of assembly, including penetrations, flashings, transitions, etc.	0.04 cfm/sq. ft. at 75 Pa (0.3" water column)
	ASTM E1677, *Standard Specification for Air Barrier (AB) Material or System for Low-Rise Framed Building Walls*[7]	Assembly lab test on opaque assembly: air leakage (E283), structural loading (E330), water resistance (E331), and water vapor permeance (E96)	0.04 cfm/sq. ft. at 75 Pa
	ASTM E283, *Standard Test Method for Determining Rate of Air Leakage through Exterior Windows, Curtain Walls, and Doors under Specified Pressure Differences across the Specimen*[9]	Assembly lab test on representative wall sample	0.04 cfm/sq. ft. at 75 Pa (0.3" water column) or as listed in Table 405.2 for fenestration assemblies
3	ASTM E779, *Standard Test Method for Determining Air Leakage Rate by Fan Pressurization*[8]	In situ test	0.4 cfm/sq. ft. at 75 Pa (0.3" water column)

Dynamic Pressure.[13] Typically, water penetration resistance values are between 6.24 lb./sq. ft. (kg/m^2) and 15 lb./sq. ft. (kg/m^2) depending on the system type. A performance mock-up (PMU) may be required for systems with unique configurations or for untested assemblies for which a portion of the aforementioned tests will likely be part of the programs. See coordination section for mock-up considerations. Field quality-control testing may include pressurized chamber testing; ASTM E1105, *Standard Test Method for Field Determination of Water Penetration of Insulated Exterior Windows, Skylights, Doors, and Curtain Walls, by Uniform of Cyclic Static Air Pressure Difference*[14]; or dynamic testing in the form of AAMA 501.1 or AAMA 501.2, *Quality Assurance and Water Field Check of Installed Storefronts, Curtain Walls and Sloped Glazing Systems.*[15] These field tests can include the transition assemblies as part of the test specimen but require it to be written into the specification or agreed upon by the project team.

NFPA 285

The proper design of flashing and closures around the fenestration is further complicated when NFPA 285 compliance is triggered when combustible materials are used in the exterior wall assembly. If NFPA 285 compliance is required, only tested wall assemblies can be utilized, including the assembly detailing of the penetration. Often, the NFPA 285 compliant details run counter to efficient thermal and moisture management design. The detail may indicate brick returns or steel elements, which provide thermal bridges and cause discontinuity in the thermal envelope. Exceptions to NFPA 285 compliance can be provided but require attention and knowledge of the building code and may require an engineering judgment. If plastic-based (combustible) insulation or asphaltic-based AWBs are used in the cavity, then NFPA 285 compliance would be triggered. The designer should be aware of the tested assemblies available. This often requires utilization of the various manufacturers for test data.

Coordination and Design

COORDINATION

The design details should take into account the sequencing and constructability of the transition. Multiple trade contractors often participate in their construction, including glazers, AWB installers, and cladding contractors, as well as the contractors involved in the construction of the backup wall that must be erected before installation of the envelope components. Shop drawing reviews, building enclosure coordination meetings, and mock-ups (**fig. 1**) are all important steps to ensure coordination so that each trade contractor's responsibilities, the detailing, and the sequencing can be refined early in the process. Ideally, mock-ups are performed by the same contractor personnel used in the field. Laboratory mock-ups are typically only required for projects that have not already been tested or are unique in their composition. For most projects, site or in-situ mock-ups suffice. These mock-ups

FIG. 1 Example of a mock-up in which the fenestration and opaque wall assembly are included.

should include the AWB assemblies and all fenestration and cladding types for the project. All transition components should be provided to include transition membranes, flashings, and closures. If possible, performance testing should be considered and should include the field quality-control testing previously mentioned.

BUILDING ENCLOSURE COMMISSIONING

To assist in facilitating the coordination between design, construction, and subtrade contractors, the building enclosure commissioning (BECx) process can be utilized. The intent of the BECx process is to provide independent review that the owners' project requirements are met and a holistic approach to quality assurance of the building enclosure. One of the primary benefits of BECx is the consistent gathering and coordination of the project team both in design and the construction. A key part of the process is the review of the interface details such as the transitions discussed herein. The transition details are reviewed by the project team prior to construction to limit compromises of the quality, cost, or schedule. Subcontractor coordination meetings are typically included in the BECx process, which can be used to verify all components are accounted for and to review sequencing.

DESIGN PRINCIPLES

The primary design principle of any transition assembly used as part of the AWB system is to provide continuity of the control layers between systems. Therefore, transition detailing must accommodate the conditions, compatibility issues, and substrate movement between the systems. In regard to the fenestration transition specifically, the design must provide continuity of the AWB from the rough opening to the corresponding air seal and waterline of the fenestration assembly. In general, the primary air seal and waterline of the fenestration are typically one and the same but may differ depending on fenestration type. The primary fenestration components to be considered during the design should be the waterline of the assembly, the framing profiles that are used to make the connections, and the air and moisture management systems integral to the fenestrations that will be utilized. Moisture management components such as weeps should not be obstructed as part of the transition.

Location

The location of the fenestration within the wall assembly is also critical to the design of the transition assembly. The placement of the fenestration relative to the exterior wall components should be dictated by the location of the thermal break of the fenestration and the wall thermal control layer of the opaque wall, the anchorage requirements of the fenestration, and the required air and water seal locations. In rainscreen applications, where the waterline of the fenestration sits to the exterior of the AWB line, cavity closures and jamb flashings may be required to separate the fenestration from the "wet" cavity and extend the dry zone of the fenestration (**fig. 2**). Materials such as wood blocking may be used to extend the rough opening further to the exterior and aid in the AWB tie-in detailing. Similarly, if the waterline is at or behind the AWB line a trim or cladding return may be required for aesthetics to cover the cavity (UV exposure) and limit moisture infiltration to the wall.

FIG. 2 Detail showing example of jamb flashing to extend the waterline of a fenestration.

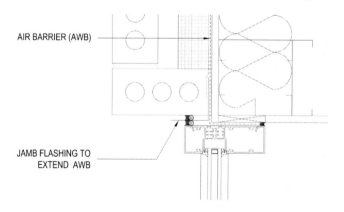

The fenestration may sit flush with the backup wall allowing for the AWB tie-ins to be straight formed.

In barrier wall assemblies, primary air and water seals are made directly between the fenestration and cladding material that serves as the AWB. Consider precast or tilt-up type barrier wall assemblies where the fenestration would be traditionally anchored directly to the concrete wall. Due to the increased concern for energy performance and awareness of risks associated with condensation, designers are moving the fenestration inward in cooler climates by adding structure to the interior of the wall. These supports, typically steel angles (**fig. 3**) or interior stud walls, may use thermal isolators between the wall and the support to separate the two components. Nonthermally conductive material such as fiberglass or wood can also be used in lieu of steel. The addition of these supports extends the rough

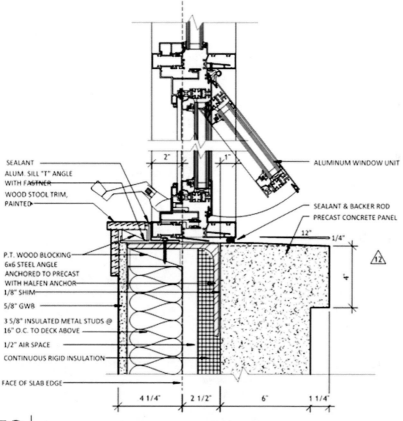

FIG. 3 Detail showing an example of interior supports for glazing systems in a barrier wall application. The detail shows a punched window sill.

opening past the inside face of the exterior wall; the AWB should also be extended to accommodate the transition seal.

Durability
Transitions should be designed to maintain continuity of the seals for the service life of the fenestration assembly. Due to the need to accommodate movement while still providing a leak-free connection, sealant and other flexible membranes are typically used as the primary transition materials. Most such materials are subject to weathering and UV degradation and may not last the service life of the fenestration proper. Therefore, durability, maintainability, and redundancy of the transition assemblies should all be considered during the design. This can include provisions for sealant replacement and the addition of secondary seals should the outer seals fail due to adhesion or general weathering.

Constructability
One of the more overlooked aspects of transition detailing is constructability. Constructability should not be judged solely on whether the detail is conceptually feasible but rather that a detail can be reasonably constructed on a large scale with labor consistent with industry standards. Fenestrations are generally repetitive in nature, so constructability issues can become systemic deficiencies. The difficulty of constructing an assembly is directly related to its components' propensity to be deficient.

Constructability also relates to sequencing and the resultant schedule consequences. The clear separation of trades, AWB versus fenestration versus cladding installers, is paramount to proper sequencing of the transition details. This separation starts with the detailing and continues through subcontractor buyout and trade coordination in the field. For example, if a jamb flashing is provided that requires the return leg into the cavity to be stripped in with AWB, then the sequencing would be as follows: AWB installation, jamb flashing installation, AWB stripping, and then the fenestration. The jamb flashing must be installed and stripped in before the cladding can be installed at these locations. Consider the same detail for which a jamb flashing is provided that returns into the rough opening with a blind seal to the AWB; the cladding could be installed prior to the jamb flashing. Construction sequencing should also be considered in the context of building "dry-in" for which interface details such as AWB tie-ins play an important role. Understanding and anticipating these construction considerations is a difficult task for designers, specifically in traditional design-bid-build projects in which dialogue on the critical path sequencing does not occur until the bidding process is complete. Ultimately, if the designer does not have some construction knowledge and forethought into these constructability and sequencing issues, the cost and schedule for the project can be affected.

Head and Sill Conditions
Much of the current literature and concentration of transition detailing centers around the jamb condition. However, the head and sill details play an important

role in rainscreen applications because they are directly involved in the drainage of the wall assembly. A common mistake when detailing sill conditions is returning a sill flashing beneath the assembly and breaking the air seal between the fenestration and the AWB. Ideally, the flashing would stop short of the thermal break of the fenestration to limit thermal bridging, terminating before the exterior air and water seal of the fenestration with the flashing edge being covered by sealant. Unfortunately, this is not usually feasible due to the desired concealment of the mechanical fasteners required to secure the flashing and related edge distance requirements. In these cases, the flashing should be set in sealant or stripped in to form a seal to the AWB. Similar concepts can be used for the jamb and head situations for which covers are used to conceal of the wall terminations.

Wall flashings are required for most rainscreen assemblies to control the moisture draining within and out of the cavity, especially before it reaches interruptions such as fenestrations. If the fenestration sits proud of the AWB line, the head flashing will serve a similar purpose as the jamb flashing of extending the AWB for the exterior seal and for closing off the cavity. Therefore, the head flashing will also need to be fully sealed to the AWB. This can be accomplished by stripping in the vertical leg of the head flashing.

In general, it is critical that the AWB tie-in be continuously sealed around the perimeter of the fenestrations. Often the points of failure are the transitions from jamb to head/sill. For hung cladding systems this can be easily achieved via a corner closure or boot between the jamb and head/sill flashing. This concept becomes more difficult in brick masonry walls with loose lintels. The jamb flashings are attached to the backup wall and will move with it; however, the lintel is embedded in the brick veneer, resulting in differential movement between the seal at the head and the jamb. There is also the issue of sequencing because the jamb flashing will have to be installed prior to the brick masonry for the return leg to be stripped in with the AWB. To account for installation tolerances, the contractor would need to leave a large gap between the lintel and the jamb flashing. To close the gap, a silicone boot could be used to bridge between the head and jamb flashings. Alternatively, a two-piece flashing could be considered in which a sleeve/cover sits on a receiver after the brick is installed and can be field cut to account for the placement of the lintel and a heel bead be used to seal the two. Similarly, there are instances in which the deflection joint of a floor runs through the head track of the window. In these scenarios prefabricated boots may be required at the three-dimensional movement corners, jamb return to head. The sealant at the head joint can typically accommodate the anticipated movement in plane.

GLAZING SYSTEMS

Two of the most common fenestration types will be further discussed as examples of fenestrations that have different detailing requirements: curtain wall and storefront assemblies. Many of the transition issues relate specifically to these systems. Designers often inappropriately treat these assemblies with similar detailing, not accounting for their unique air and moisture control systems and proper tie-in points. The fenestrations will be discussed in terms of air seals, waterline, moisture management systems, and the typical framing profiles as they relate to transition viability. Potential connection detailing for the transitions will also be discussed. It should be noted that window-type fenestrations are also commonly used and come in a variety of configurations. Aside from integral fin types often found in residential construction, the principles of curtain wall and storefront assemblies can be generally applied to window assemblies. Integral fin windows are straightforward in their detailing, with the head and jamb flashings being stripped in with AWB and bottom flanges left open, or skip sealed to allow moisture to drain in the case of infiltration to the rough opening. The continuous air seal line is provided by an interior seal to the AWB.

Curtain Wall

Curtain wall systems are non-load bearing exterior wall assemblies that span multiple floors and are typically constructed of aluminum framing with infill panels.[16] These panels can include a variety of materials such as glass, metals, composites, and so on. The term "curtain" is in reference to the system being typically hung from the building. The vertical framing members of the curtain wall, referred to as mullions, are structural in nature, resisting the lateral loading and carrying gravity loads to the structure. The horizontal members, referred to as rails, carry the gravity loads of the infill panels to the mullions via a shear connection. The lateral and gravity loads are transferred to the building via anchors that are typically located at mullions along the sill, head, and intermediate supports such as floor slabs or other framing. Many preengineered curtain wall systems are available and used throughout the building industry. More specialized curtain wall systems are developed on a project-by-project basis to accommodate various constraints, such as site limitations, aesthetics, active facade components, and increased structural or blast loading. Curtain wall systems can also be used in lieu of punched windows or storefront systems, or both (**fig. 3**).

One of the main benefits of curtain wall is its ability to accommodate glass as a continuous primary facade element. The glazing methodology may be either captured glazing or structural silicone glazing (SSG). Captured glazing systems typically rely on a pressure bar with dry gaskets and relies on compression for sealing. Wet seals can be provided. The captured glazing system is often a water-managed system in that some moisture penetration is expected and is expelled through weeps at each horizontal rail. The structural silicone glazing system utilizes silicone adhesive between the glass infill and the frame. Elastomeric face seals are provided at butt joints in the glazing; therefore, four-sided SSG systems are considered a barrier-type assembly. A combination of two-sided SSG and captured glazing in the perpendicular direction is also common.

Generally, curtain wall systems fall into two fabrication categories: stick-built and unitized. Stick-built systems are typically erected and glazed in the field. Unitized systems are typically fabricated and glazed in the shop. Unitized systems arrive on site as a finished panelized unit intended to interlock and stack. Due to the nature of the system, the unitized systems normally use four-sided SSG for glazing. Systems utilizing screw spline interlocks may be partially fabricated in the shop and can be either captured glazing or SSG. Stick-built systems typically span up to two floors and must accommodate vertical movement at the splice joint between the mullions and horizontal movement at the perimeter conditions. Anchorage occurs at the top, bottom, and floor slabs. Unitized systems accommodate movements between each panel at the interlock and stack joints. Other curtain wall systems are available such as point-supported and toggle systems. These systems are used with less regularity than the stick-built and unitized systems.

For mid-rise projects, stick-built curtain walls with captured glazing is most commonly used. As previously mentioned, captured glazing systems are designed with moisture management systems within the glazing pocket where moisture is allowed to collect and is expelled through weeps at each pressure bar or rail. The glazing pocket is the space where the glass unit sits in the framing, between the shoulder and nose of the mullion/rail. The glazing pocket is therefore considered a wet zone in the curtain wall system and is fully sealed. The waterline of a curtain wall system (**fig. 4**) is the back face of the glazing pocket or the shoulder of the frame.[17] The transition from the AWB to the curtain wall should ideally occur at this location. It should be noted that for SSG curtain wall systems the primary air

FIG. 4 Typical curtain wall with air and waterline indicated.

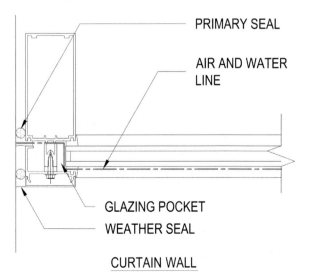

and water seal may be to the shoulder or the glazing proper because the continuous SSG seals extend the AWB to the outer face (surface #1), creating a barrier-like system.

Although there are a variety of transition methodologies, the most common connection between the AWB and curtain wall is via a continuous sealant joint between the curtain wall framing and AWB at the rough opening. This is referred to as the primary seal. The primary seal may be accompanied by a secondary seal on the interior at the back face of the mullion. This seal may provide redundancy in the air sealing of the systems and protection against moisture intrusion to the interior should the primary seal fail. Insulation of the interior seal may not be feasible on multifloor spans due to obstructions caused by structural framing and floor slabs. It is recommended that a weather seal be provided on the exterior from the mullion cap to the cladding to limit moisture to the primary seal and protect it from UV exposure. Note that the weather seal cannot be considered the air and water seal because it is proud of the wet zones of the curtain wall system. Ideally, the primary seal is installed prior to the cladding or cavity insulation to provide full access to the joint location. Tooling of the joint and quality control can be difficult if the cladding is installed beforehand. The sealant connection also requires that the AWB and backup wall are aligned so the connection can be achieved. If, for example, the curtain wall sits proud of the backup wall so that the face of cladding and face of curtain wall more closely align aesthetically, the shoulder would likely be proud of the backup wall, exposing significant portions of the "dry" mullion to a wet environment. A major disadvantage of the sealant approach is the sensitivity to changes in attachment configuration. Some curtain walls have perimeter mullions or vertical mullions with open profiles that, unlike the sill mullion, provide minimal thickness of material for the sealant to adhere to and can be a weak point. Occasionally, the jamb, head, or sill rough opening may be out of plane from each other, requiring the sealant bead to shift planes or be installed in a recessed position. Alignment of the joints around the perimeter condition is an important design aspect to review.

Where alignment issues exist or where excessive joint size/movement is anticipated, preengineered transition assemblies should be considered. These transition assemblies can be installed into the glazing pocket and can utilize the pressure bar as a compression seal. These systems incorporate membranes that must be robust and compatible with adjacent products. Examples of a typical transition membrane are preformed silicone, neoprene, PVC, and so on. The membranes should be flexible and be able to accommodate span gaps or offsets from the AWB to the glazing pocket. The membranes should also be designed to accommodate differential movement between the fenestration and the AWB. One of the most common transition membranes is preformed silicone (**figs. 5** and **6**), which has the benefit of being compatible with the perimeter weather sealant so there are no adhesion issues. The silicone sheet can be used at all perimeter conditions of the captured glazing curtain wall. There are also proprietary curtain wall systems in which a flange can be

FIG. 5 Detail indicating an extruded silicone transition membrane used in a curtain wall assembly.

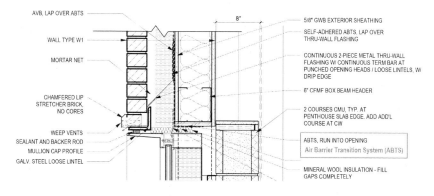

FIG. 6 Photograph of continuous silicone transition membrane at window surround. Note that a curtain wall assembly is being used in a punched window application.

inserted into a glazing pocket, allowing the curtain wall to be flashed and secured at the perimeter similar to an integral fin window unit.

Storefront

Storefront glazing assemblies are typically used in low- to mid-rise construction and in a single-story application. The assembly is also commonly used in lieu of window assemblies in a punched opening in which larger assemblies are desired. Storefront systems are typically less customized and therefore less expensive compared with curtain wall or window assemblies per unit area. Storefront framing is limited in depth, typically not exceeding 6 in., with a channel-type framing profile. The frames do not have the ability to accommodate significant reinforcement. Storefronts typically have a lower water penetration resistance rating than curtain walls. Thermally improved systems are available that utilize dual thermal breaks, but infill thicknesses are more limited; they typically cannot accommodate deep infill options such as triple-glazed insulated glass units.

Storefront assemblies can be forward, center, or back set and can be interior or exterior glazed. The system uses glass stops for installation of the glazing and is generally dry glazed, meaning the system relies on gaskets for its air and water seals. The storefront utilizes the mullions for the air and water management system (**fig. 7**), draining moisture from the glazing pocket into vertical mullions, direct by diverters, to the sill, where it is expelled via weeps at the subsill. Because the frame is open and the glass pocket is not compartmentalized in the same way curtain wall is, the waterline is technically the back face of the frame.[17] However, the primary transition to the control layers, including the AWB tie-ins, should occur at the exterior face of the storefront frame.

There are several considerations when detailing transitions of storefront assemblies to the AWB, particularly in rainscreen and barrier wall applications. As previously mentioned, the storefront frame has an open profile. The open area facing the rough opening is not intended to take on moisture (**fig. 8**) and therefore must be sealed from the wet zones of the adjacent wall and from weathering. The latter can be achieved by providing a seal at the exterior face but does not address moisture

FIG. 7 Typical storefront assembly indicating the air and waterline of the assembly.

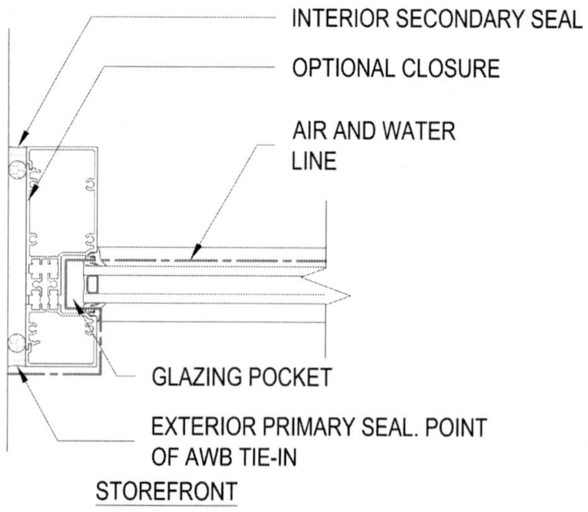

INTERIOR SECONDARY SEAL
OPTIONAL CLOSURE
AIR AND WATER LINE
GLAZING POCKET
EXTERIOR PRIMARY SEAL. POINT OF AWB TIE-IN
STOREFRONT

FIG. 8 Open framing profile of a storefront assembly. Figure depicts improper detailing of storefront assemblies that does not provide for a seal from the exterior face of the frame to the AWB.

within a cavity system. The exterior seal is also required to limit the outside air past the thermal break to reduce condensation risks. Designers should be aware that a typical storefront frame has limited depth to its return legs on the front and back of the extrusion to receive the sealant; therefore, the location of the outer and inner seals should occur only at these locations. Snap in thermal covers can be provided to limit moisture to the frame; however, these are not considered air- or watertight. Transition membranes are not typically used in storefront applications because the return legs are not deep enough to accommodate membranes such as extruded silicones, for which manufacturers usually require a 1-in.-wide minimum adhesive bed. If covers are provided, the use of transition membranes is feasible; however, the membrane would need to be run to the exterior face of the storefront. Note that transition membranes are sometimes used in lieu of, or in addition to, the interior primary seal or as secondary seal for storefront assemblies. Because of the cavity depth of current rainscreen wall assemblies combined with the relatively shallow depth of storefronts, cavity closures are typically required but often omitted from the design specifications or details and are considered optional by the manufacturer. This omission often leads to additional costs during construction. These closures should be sealed to the AWB to extend the assembly such that a continuous exterior seal can be made to the storefront frame. Similarly, in barrier wall applications in which the AWB is the outer layer of the wall, a jamb extension may be required to extend the inner seal location if the cladding does not have a deep-enough return at the rough opening. To reduce issues related to the open framing, the storefront assemblies can be set inside a receptor frame. In this scenario, the air and water seals can be made in differing locations than those previously stated. Receptor systems can also accommodate integral extensions to cover the cavity (**fig. 9**). Although the receptor system is beneficial, they can be prone to deficiencies related to workmanship because they are assembled in the field and rely on seals at the miter joints and additional gaskets for the exterior air and water seals. Additionally, if the receptor frames are not mitered, similar issues with the open-framing profiles will occur.

Case Studies

CASE STUDY 1

The subject project is a renovation and large addition to an existing primary school located near Washington, DC. The building is designed to have net zero energy consumption and is targeting LEED Gold and LEED Zero certifications. The exterior wall assemblies consisted of cold-formed steel-stud backup walls with gypsum sheathing, fluid-applied vapor-permeable AWB, 5 in. of continuous mineral wool insulation, a 2-in. air space, and various cladding materials attached via fiberglass clips to the backup wall or brick veneer. The fenestrations consisted of storefront

FIG. 9 Storefront receptor frame with integrated extensions to account for continuous insulation and cladding systems.

assemblies at entrances and in punched windows and isolated areas of curtain wall spanning multiple floors. To allow for fastening of the storefront assemblies, LVL blocking at the rough opening projected into the cavity. Similar detailing was used at curtain wall locations.

During design reviews, several concerns were noted at the fenestration openings due to the increased cavity depth. In general, the storefront systems indicated were 4.5 in. deep and were located in the wet zone of the cavity. At areas designated with brick veneer, brick returns were used to conceal the cavity (fig. 10). At areas with phenolic or metal panel cladding, phenolic panels were used at the returns. The placement of the storefront meant the exterior seal was proud of the AWB. A thermal closure for the storefront framing had been indicated by the architect, which they believed allowed the primary exterior seal to be moved back to achieve the connection to the AWB. The design was left unchanged moving into construction, where similar issues were noted during the submittal stage.

During the in-place mock-up review of head flashing detail conditions, the glazing installer reiterated similar comments raised during the design and submittal stages by the consultant: that the exterior seal must be at the face of the storefront and that the cavity must be sealed from moisture. An alternative design direction to use a cavity closure that could be integrated to the AWB was discussed, but the construction manager noted the brick veneer was the critical path for the building enclosure based on the construction schedule. The installation of additional flashing

FIG. 10 Contract document detail at brick veneer.

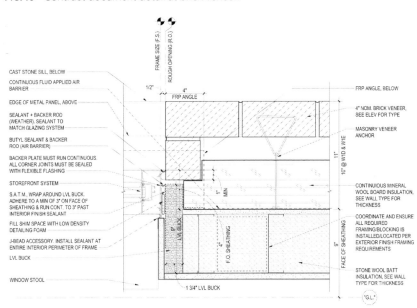

would add significant time to the enclosure schedule to accommodate mobilizations to the rough openings by multiple trade contractors and would be difficult due to limited site access for lifts. To maintain the schedule, the owner mandated that the brick veneer work move forward and the project team develop alternatives in which the cavity closures could be added after the brick. Ultimately, the two details presented included adding an extruded silicone tape between the brick and the AWB to close the cavity or provide a sealant joint between the AWB-wrapped LVL and the brick (fig. 11). The extruded silicone was a significant increase in cost compared with the sealant joint and would also add some time to the schedule. Additionally, the extruded silicone assembly would add thickness to the rough opening, estimated to be $3/8$ in., where the storefront was sized to accommodate $1/2$-in. joints at the perimeter, leaving only $1/8$ in. of tolerance. With such little room for installation, it was likely the extruded silicone could become damaged in some locations by the corners of the framing. The sealant between the brick return and the LVL closed the cavity but did not extend the AWB to the exterior seal because the brick veneer is not considered watertight. It should be noted that the AWB did return into the opening far enough for the inner seal connection to occur. Ultimately, the glazer was willing to warranty the installation using the sealant, and the owner accepted the increased risk associated with this detail.

The metal and phenolic panel cladding were not on the critical path, allowing the transitions to be provided prior to the cladding installation. The construction

FIG. 11 Revised detail showing sealant between brick and LVL to act as cavity closure.

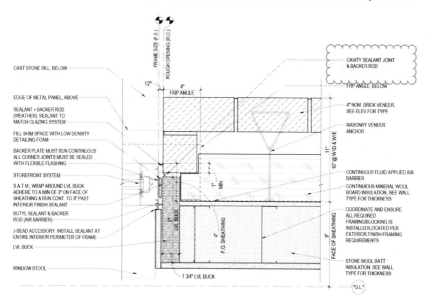

documents (fig. 12) required a phenolic panel be provided at the returns for both the metal and phenolic panel cladding (fig. 13). To accommodate the open cladding at the jamb while still extending the AWB to the exterior face, extruded silicone transition membranes were used (figs. 14 and 15). This was feasible because the snap

FIG. 12 Sealant installed between brick and AWB to close cavity.

FIG. 13 Contract document detail at phenolic panel.

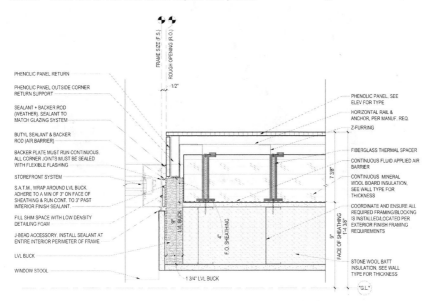

FIG. 14 Revised detail at phenolic panel indicating the extruded silicone transition assembly.

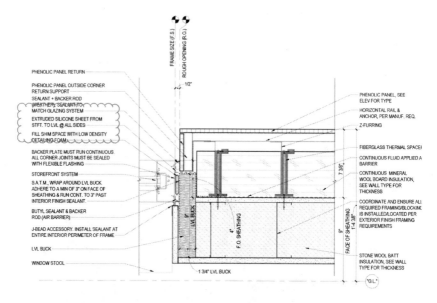

FIG. 15 Completed transition assembly.

in mullion covers provides a solid substrate for the membrane. Had these not been provided this detailing would not be possible. A cosmetic seal was provided between the phenolic panel and the storefront.

CASE STUDY 2

The subject project is a hotel and convention center in Washington, DC. The building totals over 1 million sq. ft. and includes the renovation and integration of a historic office building. The building was certified LEED Silver. The exterior walls are mainly unitized curtain wall or opaque metal-framed walls with punched windows. The opaque walls primarily consisted of cold-formed steel-stud backup walls with gypsum sheathing, vapor-retarding self-adhering AWB, extruded polystyrene insulation, and metal panel cladding on "Z-girts."

The project was designed with the typical punched window head in line with the deflection track of the backup wall. During the PMU, moisture was noted at the perimeter joint of the window when water penetration testing was performed per ASTM E331 and AAMA 501.1. The air-leakage test, per ASTM E283, was below the maximum allowable limit. Based on visual evaluation and diagnostic testing it was determined that the AWB was opening when subjected to pressure, where it turned into the rough opening at the deflection track (**fig. 16**). Multiple repair efforts and retesting were required prior to a successful test, which was dependent on the application of considerable sealant. Water penetration testing, per ASTM E1105, was performed on the first window field installations. Moisture intrusion was noted with similar failure modes at the perimeter joint as observed during the PMU testing. After multiple field-testing failures, it was determined the sealant repair was

FIG. 16 Deflection joint at the head of the rough opening.

inadequate to prevent the moisture infiltration, and the system as constructed did not meet the performance requirements. A silicone boot was proposed to accommodate the three-dimensional nature of the detail and accommodate the movement anticipated at the deflection joint while still being compatible with the perimeter sealant joints of the window (**fig. 17**). Once the boot was installed, subsequent field testing passed without water infiltration. This issue highlights the importance of a transition assembly and the surrounding AWB to be able to perform under the movement and loads anticipated. Plan and section details often do not convey the three-dimensional nature of interface conditions, requiring additional thought by the designer to convey these conditions.

CASE STUDY 3

The following study is a forensic evaluation of moisture intrusion issues being experienced at a medical facility in Baltimore, Maryland. The building is a 120,000-sq. ft. facility with five stories and a mechanical penthouse. The building was constructed between 2008 and 2010 and had been occupied since 2010. The exterior facade was constructed of masonry veneer, curtain walls, punched and ribbon window assemblies utilizing storefront framing, and metal and composite wall panel cladding over a steel and reinforced concrete structure. The building had been experiencing leaks at various locations throughout the facility for several years, particularly on the fourth and fifth floors at the northeast and northwest building corners. It was reported that several chronic leaks had persisted through multiple repair attempts. The chronic leaks prohibited these spaces from being occupiable.

FIG. 17 Silicone boot at the window head used to accommodate movement.

As part of the forensic evaluation, a review of the construction documents and metal panel shop drawing submittals was performed. A commercial building was used as the basis of design for the AWB and was submitted for the project. The submittal included installation instructions, for which the following items were of note:
- When installing as an air barrier, seal wrap the bottom of the wall with sealant, manufacturer's tape, or butyl tape.
- Secure the commercial wrap by fastening into studs. When fastening into steel studs, wrap requires 2-in. cap screws or $1^{1}/_{4}$-in. metal-gasketed washers with screws.
- For applications in which wrap is to be installed behind metal panel cladding, the manufacturer recommends butyl tape or approved alternate patches be installed behind all metal installation brackets. Window, door, and through-wall flashing shall be integrated, with the wrap ensuring proper shingling. It should be noted that the wrap is not a self-sealing membrane; therefore, any fastener holes are required to be sealed. Butyl tapes are self-sealing and would provide a seal behind the bracket.

In general, commercial wraps can be code-compliant AWB systems provided they are fully sealed per the manufacturer's instructions. Synthetic fabric weather barriers are used in rainscreen applications, particularly DBV systems such as brick veneer and siding, in which limited moisture is expected to get behind the veneer (see previous discussion regarding drained and back-ventilated rainscreen assemblies). However, synthetic fabrics are less common and not typically recommended in commercial building applications with open joint rainscreens, particularly for pressure equalized systems, where significant moisture could enter the cavity if the air barrier is not fully functional due to the considerable detailing involved in sealing penetrations, laps, and so on, in combination with wind pressures on the membrane. Self-sealing fully adhered membranes are recommended for these applications such as modified bitumen sheet applied membranes and rubberized liquid applied membranes.

Composite panels were utilized at the parapet levels and formed metal panels between rows of ribbon windows and the roof level. The composite panel had a proprietary "dry seal" system and was designed as a pressure equalized rainscreen. The formed metal wall panels were designed as a drained and back-ventilated rainscreen application. The composite metal wall panel specifications indicated that no uncontrolled water be allowed at a pressure differential of 6.24 lb./sq. ft. (kg/m^2) (75 Pa). The formed metal wall panel specification indicated no water penetration when tested according to ASTM E331 under a static pressure difference of 2.86 lb./sq. ft. (kg/m^2) and no evidence of water leakage when tested according to AAMA 501.1.

The following was of note during the review of the wall panel shop drawing submittals:
- Both the composite and formed metal panel shop drawings were marked "Approved as Noted" by the architect of record. Their commentary

primarily included notes on the AWB: "Coordinate wall panel assemblies with roofing (coping), flashing, trim, and construction of light gauge framing, soffits, and other adjoining work to provide a leakproof and noncorrosive environment." The composite wall panel submittal included additional commentary on the inspection of the water barrier by the metal wall panel installer: "Metal wall contractor shall inspect installation of vapor permeable air and water barrier for compliance with manufacturer's installation instruction. Corrections necessary shall be made prior to wall panel installation."

- A sheet metal counterflashing was to be face fastened with pop rivets to a concealed Z-girt above the window head. The Z-girt was to be fastened through the WRB to the sheathing. A butyl tape was noted at the fastener locations but was not shown to be lapped onto the flashing. This detailing was consistent with the construction documents reviewed and also consistent with the flashing at joints between the composite and formed metal wall panels.
- The counterflashing drip edge was to be hemmed over a sheet metal closure that returned horizontally into the rough opening between the window head and metal stud wall sill plate. The window sealant was to be provided from the frame to the closure. No weeps were provided in the closure.

During leak testing of the vertical joints in the composite panels along the lower formed metal panel butt joints, moisture was observed emanating from beneath the closure and commercial wrap at the interior of the window head (**figs. 18** and **19**).

FIG. 18 Moisture intrusion noted at the window head after a rain event. Note the moisture is beneath the AWB membrane as result of improper installation.

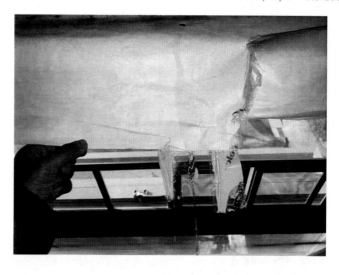

FIG. 19 Detail from metal panel shop drawing moisture path to the interior. Note the closure returning into the opening and the improperly lapped flashing at the base of the panel.

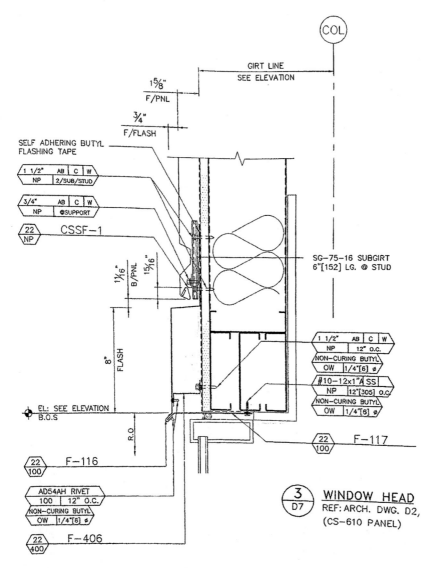

Further testing showed moisture collecting in the head of the window, eventually coming through the frame joinery. Following removal of the composite metal wall panels, it was observed that the joint interlocks were improperly installed and were allowing significant moisture into the system. Additionally, the commercial wrap was fastened with staples, not the required self-sealing fasteners. The staples and laps in the commercial wrap were not taped or otherwise sealed. As constructed the commercial wrap could not resist wind loading and therefore could not be considered a continuous air barrier. These deficiencies negated the pressure equalization design of the metal panel system above the windows. The moisture, which bypassed the wall panel and freely migrated down the face of the backup wall, was directed into the building by improperly lapped wall flashing at the base of the metal panel. The head closure returned into the rough opening of the window, bypassing the primary window seal. Additionally, the improperly lapped and sealed AWB allowed moisture behind the membrane and provided a direct moisture path to the interior. This would be true even if the flashing were lapped properly.

This forensic case study exemplifies the dependence of fenestration transitions on correct AWB installation and detailing to provide continuous control layers. In this example the commercial wrap was not installed in such a manner that provided a continuous air barrier membrane due to the unsealed penetrations and laps. Additionally, the flashings were not positively lapped to direct moisture out of the wall assembly. A head closure was used to cover the air space but was returned past the seal of the window, which allowed moisture intrusion to the window head and interior.

Summary

The integration of fenestrations in an opaque wall assembly creates unique and challenging detailing requirements for the designer. Opaque walls contain air and moisture barriers that are required to be continuous by code and for the wall to perform as intended. These layers limit air leakage from the exterior to interior and vice versa and provide a control plane for the moisture within the wall. The transition details should be fully developed and vetted prior to construction. Too frequently, these details are left at a schematic level through the construction phase, leading to modifications and additional costs. At the construction stage, compromises are often made due to scheduling and cost constraints. The multiple trade-offs involved in transition details can make field coordination and quality control difficult, resulting in deficiencies and corresponding performance issues.

To deliver transition details that provide full continuity through the fenestration, the designer should consider the type and configuration of the adjacent opaque wall assembly, the fenestration type, and where the control layers are within those assemblies. In barrier wall assemblies, the air and moisture control layers are at the exterior face of the cladding system. This can cause complications with the location of the air and water seals. Jamb extensions may be required to accommodate the fenestration systems, particularly when used with storefront assemblies. The increased use of rainscreen wall assemblies has complicated wall detailing immensely for today's designer. The rainscreen wall must accommodate cladding, air space, thermal insulation, and AWB membranes. When integrating a fenestration into these wall assemblies, careful detailing must be used so as not to allow the air or moisture in the drainable cavity behind the corresponding seals to dry zones in the fenestration. The fenestration placement should also be coordinated with the placement of the wall and insulation and thermal break of that system. In many cases, covers and closures are required to extend the air and water barriers further to the exterior. This is to accommodate the exterior seals and so the fenestration is fully segregated from the cavity.

Curtain wall and storefront assemblies were used as examples of systems that have unique detailing requirements of the air seal and waterlines that are typically misunderstood by designers. The primary air seal and waterline of a curtain wall occurs at the shoulder of the mullion; this is where the AWB connection should occur. This connection can be accomplished by a sealant joint or a transition membrane assembly. Storefront assemblies are commonly used in commercial construction for their versatility and cost. Due to the open nature of the framing profiles and the moisture management system being in the mullion, they often require additional components be provided to extend the AWB to provide connections at the exterior face of the assembly.

The case studies provided show that transitions should be reviewed and coordinated with the project team in the design phase to limit issues during construction such as sequencing, schedule and cost impacts, and quality control. These issues can lead to deficiencies in the transitions, resulting in air leakage or moisture infiltration. Case Study 1 discussed a project in which the jamb closures were not provided, and the exterior seal for the storefront assembly could not be installed on the rainscreen cladding as shown. Multiple options were discussed, but ultimately a compromise was made by using a sealant joint between the brick return and the AWB to seal the cavity. Case Study 2 showed design considerations for AWB transitions such as differential movement between the fenestration and the opaque wall. Case Study 3 is an example of how improper detailing of closures, specifically when returned into the opening, can result in a breach of the air and water control layers to the fenestration. In general, if the AWB is not continuous due to installation deficiencies, issues with transition detailing at the fenestration can be futile because moisture could bypass the transition seals.

References

1. L. Anastasi, G. Eckhardt, and B. Neely, "Specifying Air Barriers," Technical Paper No. 1, Construction Specifiers Institute Boston, http://web.archive.org/web/20210322142103/https://www.generalinsulation.com/wp-content/uploads/2014/07/CSI-Boston-Technical-Paper-No-1-Specifying-Air-Barriers-2.pdf
2. International Code Council, *International Building Code* (Falls Church, VA: Author, 2018).
3. International Code Council, *International Energy Conservation Code* (Falls Church, VA: Author, 2018).

4. *Standard Fire Test Method for Evaluation of Fire Propagation Characteristics of Exterior Wall Assemblies Containing Combustible Components*, NFPA 285 (Washington, DC: National Fire Protection Association, 2012).
5. *Standard Test Method for Determining Air Leakage Rate and Calculation of Air Permeance of Building Materials*, ASTM E2178-21 (West Conshohocken, PA: ASTM International, approved June 1, 2021), https://doi.org/10.1520/E2178-21
6. *Standard Test Method for Determining Air Leakage Rate of Air Barrier Assemblies*, ASTM E2357-18 (West Conshohocken, PA: ASTM International, approved October 1, 2018), https://doi.org/10.1520/E2357-18
7. *Standard Specification for Air Barrier (AB) Material or Assemblies for Low-Rise Framed Building Walls*, ASTM E1677-19 (West Conshohocken, PA: ASTM International, approved July 1, 2019), https://doi.org/10.1520/E1677-19
8. *Standard Test Method for Determining Air Leakage Rate by Fan Pressurization*, ASTM E779-19 (West Conshohocken, PA: ASTM International, approved January 1, 2019), https://doi.org/10.1520/E0779-19
9. *Standard Test Method for Determining Rate of Air Leakage through Exterior Windows, Skylights, Curtain Walls, and Doors under Specified Pressure Differences across the Specimen*, ASTM E283/E283M-19 (West Conshohocken, PA: ASTM International, approved August 1, 2019), https://doi.org/10.1520/E0283_E0283M-19
10. *Standard Test Method for Field Measurement of Air Leakage through Installed Exterior Windows and Doors*, ASTM E783-02(2018) (West Conshohocken, PA: ASTM International, approved October 1, 2018), https://doi.org/10.1520/E0783-02R18
11. *Standard Test Method for Water Penetration of Exterior Windows, Skylights, Doors, and Curtain Walls by Uniform Static Air Pressure Difference*, ASTM E331-00(2016) (West Conshohocken, PA: ASTM International, approved August 1, 2016), https://doi.org/10.1520/E0331-00R16
12. *Standard Test Method for Water Penetration of Exterior Windows, Skylights, Doors, and Curtain Walls by Cyclic Static Air Pressure Difference*, ASTM E547-00(2016) (West Conshohocken, PA: ASTM International, approved August 1, 2016), https://doi.org/10.1520/E0547-00R16
13. *Water Penetration of Windows, Curtain Walls and Doors Using Dynamic Pressure*, AAMA 501.1-17, (Schaumburg, IL: AAMA, 2017).
14. *Standard Test Method for Field Determination of Water Penetration of Installed Exterior Windows, Skylights, Doors, and Curtain Walls, by Uniform or Cyclic Static Air Pressure Difference*, ASTM E1105-15 (West Conshohocken, PA: ASTM International, approved August 1, 2015), https://doi.org/10.1520/E1105-15
15. *Quality Assurance and Water Field Check of Installed Storefronts, Curtain Walls and Sloped Glazing Systems*, AAMA 501.2-15 (Schaumburg, IL: AAMA, 2017).
16. N. Vigener and M. A. Brown, "Curtain Walls," *Whole Building Design Guide*, May 10, 2016, http://web.archive.org/web/20210322140626/https://www.wbdg.org/guides-specifications/building-envelope-design-guide/fenestration-systems/curtain-walls
17. J. Szabo and J. Sanchez, "Making Transitions to Fenestration," *Building Enclosure*, March 23, 2020, http://web.archive.org/web/20210322141438/https://www.buildingenclosureonline.com/articles/88873-making-transitions-to-fenestration

STP 1635, 2022 / available online at www.astm.org / doi: 10.1520/STP163520210047

Emily R. Hopps[1] and Anna M. Burhoe[1]

Considerations for Unitized Building Enclosure Systems: Selecting, Designing, Installing, and Testing Unitized Building Enclosure Systems

Citation

E. R. Hopps and A. M. Burhoe, "Considerations for Unitized Building Enclosure Systems: Selecting, Designing, Installing, and Testing Unitized Building Enclosure Systems," in *Building Science and the Physics of Building Enclosure Performance: 2nd Volume*, ed. D. J. Lemieux and J. Keegan (West Conshohocken, PA: ASTM International, 2022), 133–143. http://doi.org/10.1520/STP163520210047[2]

ABSTRACT

Expediting design and construction. Maximizing performance. Advancing technology and integration. The design and construction industry is increasingly focused on meeting higher performance standards while also reducing design and construction schedules. As a result, the use of unitized or prefabricated building enclosure systems also continues to grow. These systems may streamline construction, improve quality control, and enhance performance; however, they are not necessarily appropriate for every project. These systems require more significant coordination between the design and construction teams than typical enclosure construction methods; they often rely on challenging transitions at interfaces between different enclosure systems that cannot be accessed in the future; and, once manufactured, they cannot be easily altered to adapt to unforeseen site conditions. In this article, we discuss several popular unitized facade systems, reviewing the advantages and disadvantages of each. We explore design, constructability, coordination, durability, performance, continuity, and internal and external system integration. We review the benefits and challenges of

---

Manuscript received May 24, 2021; accepted for publication September 7, 2021.
[1]Simpson Gumpertz & Heger Inc., 480 Totten Pond Rd., Waltham, MA 02451, USA
[2]ASTM Second Symposium on *Building Science and the Physics of Building Enclosure Performance* on April 24–25, 2022, and June 12, 2022 in Seattle, WA, USA.

Copyright © 2022 by ASTM International, 100 Barr Harbor Drive, PO Box C700, West Conshohocken, PA 19428-2959.

ASTM International is not responsible, as a body, for the statements and opinions expressed in this paper. ASTM International does not endorse any products represented in this paper.

unitizing the enclosure systems. Finally, we summarize various techniques and approaches to quality control during the design, fabrication, and installation phases of prefabricated building enclosure systems.

Keywords

unitized panels, unitized building enclosure systems, enclosure, prefabricated panels, prefabricated building enclosure systems, megapanels

Introduction

As with almost all design and construction practices, there is an optimal time and place for the use of unitized building enclosure systems. These systems lend themselves to streamlining construction, improving quality control (QC), and enhancing performance; however, they are not necessarily appropriate for every project. Throughout this paper we highlight characteristics of unitized enclosure systems that can enhance or add challenges to enclosure design and construction to help project teams determine whether a unitized assembly will be beneficial for their project. Additionally, we summarize the various types of unitized enclosure systems, the benefits and challenges of each of these systems, and critical considerations during the design and construction of these assemblies.

What Is a Unitized Building Enclosure System?

A unitized building enclosure system is generally composed of prefabricated portions or panels of exterior wall assemblies, with each panel including some or all of the primary elements of the exterior wall: exterior cladding and attachments, fenestration, air/water/vapor and thermal control layers, and structural framing. These panels are prefabricated in a controlled factory setting, shipped to the site, and erected panel by panel. Panels are often referred to as unitized panels, prefabricated panels, or megapanels and can potentially include most types of exterior wall systems, such as curtain wall, masonry or metal panel veneers, precast concrete, and so on, both with and without integrated punched fenestration systems.

When Are Unitized Building Enclosure Systems Most Beneficial?

Unitized building enclosure systems are most beneficial for project sites without significant on-site laydown areas or challenging exterior access conditions, for projects with accelerated construction schedules, when construction is anticipated during adverse environmental conditions, when the building enclosure is repetitive, and when panels can be fabricated and transported to the site economically.

Unitized panels are not ideal for every project. Unitizing the enclosure has some potential disadvantages, such as longer design and preconstruction phases, higher upfront engineering costs, and complicated design and construction coordination.

The most challenging application of unitized panels is on a building with multiple unique wall conditions, transitions, and complex geometry, which can slow down the overall design and construction process and present a higher potential for enclosure-related performance issues. For buildings with nonrepetitive complex geometry, traditional field-fabricated or "stick-built" exterior wall systems may be a better option unless other project conditions, such as limited laydown space or shortened construction schedules, outweigh the challenges associated with complex building geometry. For a project for which a unitized enclosure will be used, below we discuss challenges and considerations for various panelization strategies.

Unitized Panel Elements

As noted above, each unitized panel includes some or all of the primary elements of the exterior wall. Depending on the type of exterior wall system, the primary elements of the exterior wall will vary; however, in general, they will include:
- Structural framing: The structure of the unitized panel can be composed of precast concrete (i.e., self-supporting), wood framing, or cold-formed metal framing. Metal-framed systems are most common for commercial construction. Similar to stick-built construction, unitized panels need to be designed to accommodate anticipated exterior wall dead loads and lateral loads; however, unlike stick-built construction, unitized panels must also withstand lifting loads during transportation and installation and thus require more rigidity to prevent racking. As such, the panel perimeter framing and interior studs are typically more robust than standard stick-built construction. The frame for each panel also includes connections to the building structure and components to facilitate integration between panels. These interpanel joints must be designed to accommodate in-plane floor slab deflections and maintain continuity of the air/water/vapor control layers.
- Exterior enclosure components: Typical exterior wall system elements include cladding, the air/water/vapor control layer system, insulation, fenestration, and cladding attachments (**fig. 1**). Most often, unitized panels are either a rainscreen system or unitized curtain wall. Rainscreen systems typically include a dedicated air/water/vapor control layer, continuous exterior insulation, a drainage cavity, and exterior veneer (e.g., metal panels, masonry) with associated attachment. The panel may also incorporate fenestration in punched openings or intermittent through-wall flashing, or both, to promote drainage of water within the panel cavity to the building exterior. The design of the exterior enclosure components is similar to typical stick-built enclosures; however, because these systems are constructed in a controlled factory setting, there is potential for better quality assurance (QA) and QC of the installation workmanship. Additionally, the detailing at panel joints and transitions between the panels and adjacent enclosure systems differs from stick-built construction, making selection and design of the panel joints critical to long-term performance.

FIG. 1 Typical exterior wall components of a rainscreen wall.

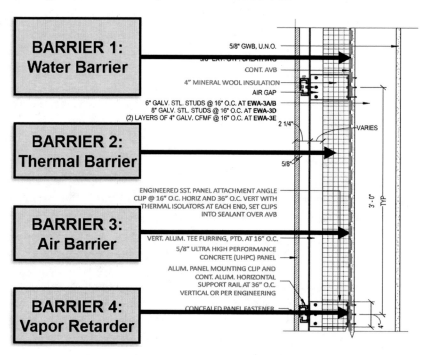

Unitized Panel Joint Options

A critical decision for unitized panels is the type of panel-to-panel joint that will be utilized. Not all unitized enclosure system joint options are equal, and the joint selection is a primary consideration affecting the long-term performance of the exterior wall system. The continuity of the exterior wall control layers (air, water, thermal, and vapor) must be maintained across these joints, which must also accommodate anticipated building and material movement. These joints are often difficult if not impossible to access and maintain after installation, so the long-term durability and reliability of the joints are crucial.

There are several types of unitized panel joints: single or dual sealant joints, membrane-covered joints, precompressed foam tapes, rubber gaskets, and interlocking perimeter frame extrusions with rubber gaskets. These joint types can be divided into two categories: barrier systems that do not include internal water management provisions inboard of the exterior surface (e.g., single sealant joints and expandable foam tape) and rainscreen systems that collect and drain water that penetrates the exterior cladding. Joints with redundancy and the ability to collect and drain water are typically more reliable and durable than those without.

- Dual sealant joints: Dual-stage sealant joints are most common on precast concrete panel walls. Dual sealant joints consist of two beads of field-installed sealant with weeps in the exterior bead to allow water that migrates inboard of the exterior seal to drain (**fig. 2**). Some systems only use a single-stage sealant joint; however, this does not provide redundancy and will rely on ongoing maintenance to remain weathertight. Dual-stage sealant joints are generally easy to install, cost-effective, and provide redundancy. However, the sealant joints must be well adhered to adjacent surfaces, continuous at intersecting panel joints, and able to accommodate any anticipated building and material movement to be effective, and because the ability to access or even view the recessed primary seal in the field is limited, it can be difficult if not impossible to confirm continuity, adhesion, and cure and impossible to perform required periodic maintenance and replacement without removal of the outboard weather seal. It is important to note that accessing the panels once they are on the building to complete the joint detailing

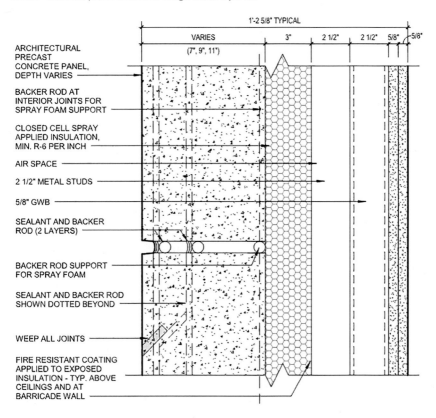

FIG. 2 Unitized panel with dual-stage sealant joints.

negates some of the most significant benefits of using the panelized system.
- Membrane-covered joints: To improve the long-term durability and reliability of unitized panel joints compared with dual sealant joints, a continuous, adhered membrane can be used to cover the joint. This membrane may be a precured silicone sheet bed in compatible sealant or a self-adhered rubberized asphalt membrane installed onto the air/water/vapor control layer of adjacent panels, spanning the joint with design provisions to accommodate movement (**fig. 3**). Like sealants, these membranes are common and relatively cost-effective, and they provide enhanced reliability compared to the sealant option because the membrane can better accommodate movement and requires less maintenance. The disadvantage of membrane-covered joints is that access to the weather control layer on adjacent panels is required to install the membrane, and therefore the exterior cladding and insulation at panel edges must be installed in the field, not in the fabrication shops. This can lead to more visible cladding joints. Similar to dual-stage sealant joints, membrane-covered joints also require accessing the panels once they are on the building to complete the joint detailing, which negates some of the most significant benefits of using the panelized system.
- Precompressed foam tapes: Unitized panel joints may also be sealed with silicone-coated precompressed foam tape. The tape is adhered to a panel prior

FIG. 3 Unitized panel with membrane-covered joint.

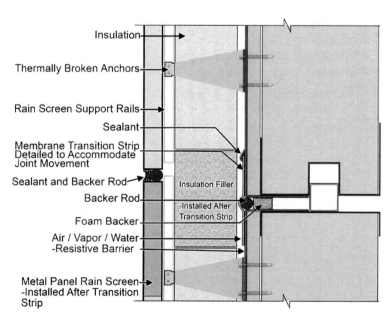

to installation of the adjacent panel and expands to create a compression seal once the adjacent panel is installed (**fig. 4**). Like sealants, precompressed foam tapes are relatively quick to install and cost-effective; however, unlike the membrane, they do not require the exterior cladding and insulation along the joints to be installed in the field. They also provide some limited thermal continuity at the joints, and the compression joint limits reliance on maintenance-intensive seals between the panels. The vapor permeability of foam tapes must be considered in the system design, and detailing at uncoated tape ends may require additional sealant application in the field at the time of panel installation. These foam tapes are workmanship-sensitive and require continuous, even, and solid surfaces to achieve adequate bond and compression. If the panel edges are not fabricated as such or if they are installed with some variation, the foam tape could roll, bulge, or tent, creating discontinuities in the weather barrier. Like recessed sealant joints, the field inspection and QC of precompressed foam tape is typically challenging, and it can be difficult if not impossible to visually confirm continuity after panel installation.

- Rubber gasket joints and interlocking frame joints with rubber gaskets: Unitized panel joints may also be detailed with rubber gaskets. Gaskets, like foam tape, are relatively quick to install and can be preinstalled in the fabrication shop. They rely on interlocking and compression between gaskets to create continuity of the air and water control layers for the building. Like precompressed foam tape joints, achieving continuity of gaskets at gasket joints and intersections can be difficult and may require field-applied sealants. Gaskets can also age and become brittle over time and can be difficult to maintain and replace.

FIG. 4 Unitized panel with precompressed foam joint.

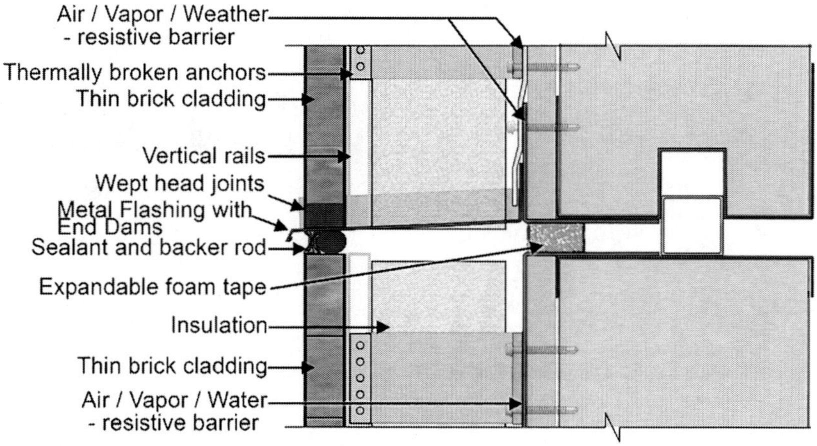

The most reliable unitized panel joint option uses continuous, interlocking male-female extrusions at panel perimeters with rubber gaskets and membrane flashing. These joints require more complex engineering and are typically more expensive than the previously described options; however, they can increase installation speed and include redundancy and internal drainage. These systems can include two rows of gaskets on vertical projections in the perimeter extrusion, often referred to as a "double chicken head," at horizontal or "stack" joints (**fig. 5**). The membrane is installed between panels at stack joints to provide continuity and to compartmentalize the system. To reduce costs, vertical joints, or mating mullions, are sometimes detailed with a single row of gaskets, but dual gaskets offer more redundancy (**fig. 5**). These panels require continuous metal perimeter extrusions, which are conductive and may act as a thermal bridge, potentially reducing the overall panel thermal performance and increasing the potential for localized interior condensation. As such, extrusions are typically designed with continuous thermal

FIG. 5 Unitized panel with interlocking rubber gaskets.

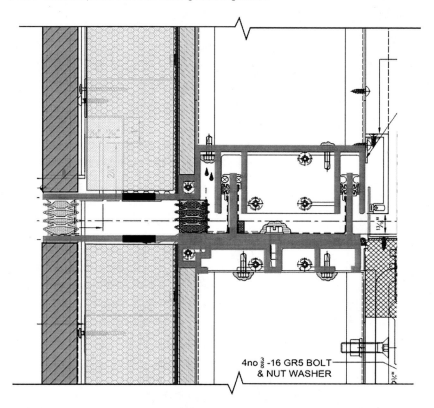

breaks and evaluated for potential condensation under project-specific interior and exterior design conditions.

Thermal Performance Design

Unitized panel construction must meet the same building code thermal performance requirements as stick-built assemblies. Opaque panelized wall systems can typically be designed to prescriptive energy code requirements and may include interior and exterior insulation, but thermal continuity can be difficult at panel joints and perimeters. The interior and exterior insulation is frequently discontinuous at panel joints, and insulation is interrupted by structural elements adjacent to joints. There are many methods for improving thermal continuity and reducing potential for condensation, such as using thermally broken extrusions, forming extrusions of lower-conductivity materials, and incorporating structural thermal breaks (**fig. 6**), but it is typically necessary to perform computer modeling to assess both thermal performance and condensation potential.

Coordination with Other Building Enclosure Components

Integration between the exterior wall components and adjacent enclosure components, such as roofs and below-grade waterproofing, can be more challenging with panelized systems and requires more attention during installation because field adjustments to unitized panels are not practical. The integration detailing

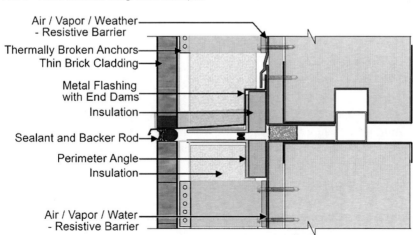

FIG. 6 Thermal break integration example.

between the unitized panels and adjacent construction should be coordinated prior to panel fabrication, which is frequently before the subcontractor trades that will be doing the adjacent work are integrated into the project team. One potential solution to reduce the preconstruction coordination required is to stick-build the first floor and any floors above the roof line because these levels tend to include many unique elements and transitions that benefit from the ability to make adjustments in the field, which is not practical with unitized panels. Components that penetrate or are attached to the enclosure, such as plumbing and electrical penetrations and interior sheathing, which are not typically coordinated with the exterior wall construction, need to be fully coordinated prior to panel fabrication. Early planning and coordination are critical to long-term performance of unitized enclosures.

QC Procedures

Unlike standard stick-built construction, unitized panel construction should include detailed QC procedures during fabrication, in addition to typical on-site construction and postinstallation QC and performance testing, to avoid systemic performance issues due to lack of familiarity with the specified system or materials, proper training for mechanics, and QC supervisor oversight. Because the panels are constructed in a controlled facility with all installers in one place, it is easier to establish a QC procedure to verify appropriate detailing and workmanship. Typically, a tag or QA/QC checklist log will be attached to each panel and, as the component or detail is installed, the tag will be checked off and signed, helping to track progress and verify completeness. These protocols typically include regular material and machine performance verification tests, such as sealant ratio and adhesion tests, de-glazing tests, and membrane adhesion tests. The project team should visit the fabrication facility early in the fabrication process to verify the understanding of the system and the implementation of the established QC protocol and should continue to visit throughout the fabrication process to review the ongoing construction.

Additionally, because these panels are custom-designed for each project and the typical panel design is repeated around the building, it is important to include a preconstruction panel assembly performance mock-up prior to wholesale fabrication to further reduce the potential for systemic performance issues during construction, potentially resulting in significant remedial work and scheduling ramifications. At a minimum, the test specimen should include, as applicable to the project, multiple panels and panel joint configurations (e.g., horizontal and vertical joints; two-way, three-way, and four-way joints) and unique fenestration systems. Mock-ups are typically constructed and tested in a laboratory or at the panel fabrication shop. Typically, performance mock-up tests include a series of structural, air-infiltration, and water-penetration tests.

Summary of the Advantages and Challenges of Unitized Building Enclosure Systems

The simplified pros and cons of using unitized enclosure systems are as follows:

Typical advantages:	Typical challenges:
• Expedited enclosure construction. • Requires less laydown area and does not require staging or multiple aerial lifts. • The majority of the enclosure materials are installed in a factory setting for improved QA and QC. This is also beneficial when construction is anticipated during adverse environmental conditions. • Sole-source construction responsibility for the majority of the enclosure.	• Longer design and preconstruction phase. • Higher upfront engineering costs. • Coordination with other trades and systems through the design-assist process. • Panel joint design and construction are critical to building performance. • Maintenance of primary weather control layers is challenging at joints. • Transitions to adjacent enclosure elements and difficulty of on-site panel adjustments. • Potential for systemic performance issues to arise if system-specific QA/QC protocol is not established and implemented early in the project.

STP 1635, 2022 / available online at www.astm.org / doi: 10.1520/STP163520210054

Andrew A. Dunlap[1] and Ryan Asava[1]

Three-Dimensional Thermal and Condensation Risk Analysis of Cantilevered Curtain Wall Elements

Citation

A. A. Dunlap and R. Asava, "Three-Dimensional Thermal and Condensation Risk Analysis of Cantilevered Curtain Wall Elements," in *Building Science and the Physics of Building Enclosure Performance: 2nd Volume*, ed. D. J. Lemieux and J. Keegan (West Conshohocken, PA: ASTM International, 2022), 144–173. http://doi.org/10.1520/STP163520210054[2]

ABSTRACT

Current design trends often result in extending curtain wall systems beyond the thermal boundary of the exterior enclosure. A common example of this condition is where a curtain wall system extends above the roof and becomes the parapet. Other examples of this type of situation include where the curtain wall extends past a floor line to create a soffit enclosure or past an adjacent perpendicular wall to create what is often referred to as a "wing wall." These types of conditions are not considered in the standard test methods that define the system's thermal performance and condensation resistance and can often lead to installations with increased heat loss and reduced condensation resistance. These types of conditions may deviate from the published performance of the system and impact the overall energy efficiency of the building enclosure. The current industry standard computer modeling methods to determine the thermal performance and condensation resistance only include two-dimensional computer modeling. Although this may be sufficient for the performance in the field of a curtain wall system, it may not adequately address these unique

---

Manuscript received August 3, 2021; accepted for publication November 2, 2021.
[1]Building Technology Studio, SmithGroup, Inc., 500 Griswold St., Ste. 1700, Detroit, MI 48226, USA A. A. D. http://orcid.org/0000-0001-8590-6458, R. A. http://orcid.org/0000-0001-7079-3913
[2]ASTM Second Symposium on *Building Science and the Physics of Building Enclosure Performance* on April 24–25, 2022, and June 12, 2022 in Seattle, WA, USA.

Copyright © 2022 by ASTM International, 100 Barr Harbor Drive, PO Box C700, West Conshohocken, PA 19428-2959.

ASTM International is not responsible, as a body, for the statements and opinions expressed in this paper. ASTM International does not endorse any products represented in this paper.

conditions. The two-dimensional nature of the evaluation cannot include the linear transfer of heat flow in the third dimension through the mullions extending beyond the thermal envelope. Laboratory testing of these complex project-specific conditions is often impractical due to cost and schedule. Additionally, once the condition is installed it is often too late to change the design. This paper evaluates the three-dimensional effects of heat flow through curtain walls using three-dimensional modeling. The relative impact on thermal performance and condensation resistance is compared with the published performance of standard curtain wall systems. The paper also reviews installation methods to decrease the heat loss and produce more resilient systems. Two-dimensional computer modeling is utilized to compare the results with the three-dimensional modeling procedures.

Keywords
condensation, curtain wall systems, high humidity, condensation control, hygrothermal analysis, three-dimensional (3D) thermal analysis

Background and Introduction

It is recognized that thermal bridging in building enclosure components, assemblies, and systems can contribute to reduced thermal performance of buildings and potentially can result in the increased risk of condensation formation. For fenestration products, the thermal performance is typically represented by the total product assembly U-factor, and common condensation-related metrics such as the condensation resistance factor (CRF)[1] and condensation index (CI)[2] are often referenced. The demonstration of these performance metrics related to fenestration assemblies such as windows, storefronts, and curtain walls has been somewhat limited in the past. However, multiple methods have been developed within the glass and glazing industry to predict product performance of fenestration systems. Current methods of evaluating fenestration system typically include two-dimensional finite element computer modeling and in some cases laboratory testing. Although there have been some advances in the recent past related to computer-simulated thermal modeling, many of the methods are still limited to two dimensions. Physical testing is often considered the best method for accurately predicting in situ performance. However, physical testing can be limited due to the cost and the amount of time/schedule required, especially when considering project specific conditions or custom designs, or both. When testing is performed, it is generally used either to validate computer modeling for general product performance ratings or for custom glazing assemblies.

ASTM C1363, *Standard Test Method for Thermal Performance of Building Materials and Envelope Assemblies by Means of a Hot Box Apparatus*,[3] and AAMA 1503, *Voluntary Test Method for Thermal Transmittance and Condensation Resistance of Windows, Doors, and Glazed Wall Sections*,[1] are laboratory test methods that are used to provide total product thermal performance (U-factor). ANSI/NFRC 100,

Procedure for Determining Fenestration Product U-Factors,[4] is commonly used to simulate the total project thermal performance using two-dimensional finite element modeling.

AAMA 1503, often referred to as the CRF test, is also used to determine the condensation resistance of fenestration systems. ANSI/NFRC 500, *Procedure for Determining Fenestration Product Condensation Index Ratings*,[2] is another method for evaluating the condensation resistance of fenestration assemblies through the use of finite element modeling.

These test and modeling methods are very useful when comparing the performance of fenestration products. However, there are limitations to these methods that do not lend themselves to determining the performance of fenestration systems installed in a cantilevered condition such as where a curtain wall extends past a roof line and functions as a parapet. The following are a few examples of the limitations when attempting to evaluate unique conditions such as parapets:

- The methods only allow for an assembly with two sides, and interior side and an exterior side. These are essentially planar specimens intended to be placed within the field of a wall. They do not include extensions where a portion of the assembly is exposed to the exterior on both sides (similar to a parapet condition) and where another portion of the assembly has a conditioned interior space on one side and the exterior on the other.
- The test and simulation methods are often based on industry standard interior and exterior temperature conditions. Although variations of the conditions can be performed, they are often not published in the manufacturers' standard literature.
- The standard methods typically do not include spandrel areas that are common at cantilevered curtain wall conditions and in particular at parapet conditions.
- The methods are based on standardized specimen configurations, dimensions, and mullion patterns that are not necessarily applicable to a specific design.

As an example, **figure 1** is a typical configuration of an AAMA 1503 test. As indicated, this type of test is simplified to a basic configuration that is exposed to exterior conditions on one side and interior conditions on the other. Even if a manufacturer elected to include items such as insulated spandrel conditions or multiple mullion patterns during the physical testing, it is costly and is only accurate at one set of interior and exterior temperatures per test. Testing several variations often becomes cost-prohibitive and can be time-consuming.

Additionally, when considering the use of industry standard computer analysis methods, heat transfer is typically considered to flow from a warm environment to a cold environment through the frame and insulating glass units in the direction perpendicular to plane of the fenestration product. This type of analysis generally considers the heat flow in two dimensions and does not consider the in-plane flow of heat through the fenestration system components, such as along the length of the

FIG. 1 Typical AAMA 1503 glazed wall system test configuration.

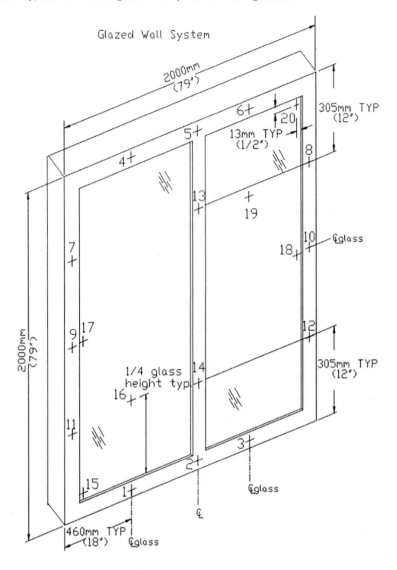

mullions. Typical two-dimensional computer modeling does not consider the complex three-dimensional interfaces at intersections of the vertical and horizontal mullions. Most modeling procedures evaluate the cross section of the fenestration product and sometimes include transitions to adjacent enclosure assemblies. However, when evaluating the results of physical test results, one can see that the interface of the various components can have an impact on the resultant interior surface temperatures compared with similar conditions at the midspan of the mullions that are impacted as much by the adjacent mullions (**fig. 2**).

Just as these differing results can been seen at the interfaces of the tested window in **figure 3**, similar but potentially more complex conditions can be present when using curtain wall constructions where they interface with adjacent exterior enclosure assemblies such as roofs, soffits, or adjacent walls. Curtain wall assemblies are typically hung on the outside of a building's structure and when used in a parapet-type condition, they often span past the roof structure. Spandrel glass and spandrel insulation are often incorporated to conceal the structure of the roof and parapet behind the curtain wall. Similarly, curtain walls are sometimes designed extending beyond adjacent perimeter wall assemblies (sometimes referred to as wing walls) or past a floor line of an overhanging soffit condition. As a result, the curtain wall assembly is exposed to the exterior condition on the front and back surfaces where it extends beyond the adjacent construction.

FIG. 2 Example results of a physical thermal performance test. Note variations at the interfaces of vertical/horizontal mullions compared with similar conditions at the midspan of the mullion.

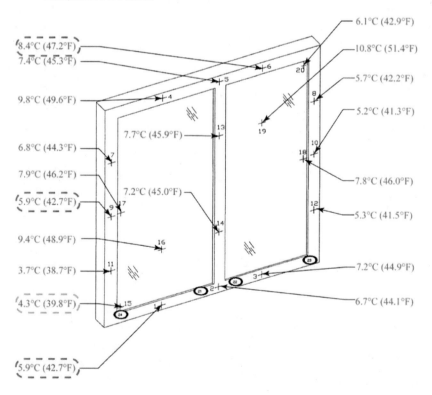

FIG. 3 Examples of curtain wall wing wall at adjacent wall, soffit, and parapet.

The remaining portion is exposed to the interior conditions on the back surface only.

In these types of conditions, the vertical mullions (and sometimes the horizontal mullions) span from the vision area into the adjacent exterior enclosure assembly. This component initially starts out being exposed on one side to the interior and the other to the exterior. However, when it extends into or past the adjacent enclosure assembly, it is then exposed or partially exposed to an exterior environment on both sides. This paper primarily focuses on a curtain wall that extends past a roof line and becomes part of the parapet construction. As the vertical mullions extend from a warmer conditioned interior environment into the colder parapet construction, the mullions are essentially vertical thermal bridges (**fig. 4**).

The overall influence of the thermal bridge on the condensation resistance and thermal performance will vary depending on several factors related to both the curtain wall construction and the parapet construction. Variables related to curtain wall include, for example, the type of vision or spandrel glazing (single pane, double pane, triple pane), or both; insulating glazing unit (IGU) spacer type; whether or not spandrel insulation was used; spandrel insulation thickness; location of the insulation within the curtain wall assembly; curtain wall vapor barrier methodology (foil-faced insulation or metal backpan); curtain wall assembly type (stick-built or unitized); curtain wall framing type (thermal broken, thermally improved, nonthermal); location of horizontal mullions relative to the roof structure and roof insulation; and types of horizontal mullions (none, fixed, or stack mullions). Other variables not specific to the curtain wall include whether a fire-rated joint assembly is utilized, location of the fire-rated joint assembly relative to the roof structure and roof insulation, parapet structure construction method (cold-formed metal framing or concrete masonry unit), method of insulating the parapet construction, whether insulation is provided above the curtain wall, and height of the parapet.

FIG. 4 Common section through a curtain wall extending beyond a low-slope roof assembly and functioning as part of the parapet construction, illustrating heat flow through the vertical mullions that run into the parapet (partially exposed condition).

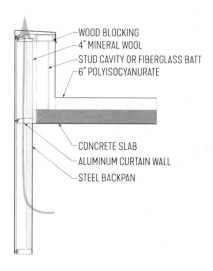

Computer Modeling Approach and Correlation

As indicated, the typical methods of analysis and testing generally do not adequately address the impacts on the performance attributes of fenestration systems when they are included in or extend beyond adjacent enclosure assemblies. This paper continues with an exploration of the use of three-dimensional finite element thermal modeling to evaluate these complex interfaces. The following is a description of the software that was used for this evaluation of common curtain wall parapet conditions and the comparison of results from two-dimensional and three-dimensional software.

THERM 7.5/WINDOW 7.5

As described by its developer, Lawrence Berkeley National Laboratory, "THERM is a state-of-the-art, computer program for use by building component manufacturers, engineers, educators, students, architects and others interested in heat transfer. THERM models two-dimensional conductive heat-transfer effects in building components such as windows, walls, foundations, roofs and doors where thermal bridges are of concern. Heat-transfer analysis, based on the finite element method, allows for evaluation of a product or system's energy efficiency and local temperature patterns, which can help identify or may relate directly to problems with condensation, moisture damage and structural integrity." This is the type of software that is most often used when performing ANSI/NFRC 100 and ANSI/NFRC 500. These procedures uses multiple two-dimensional models to produce a "total product U-factor" and "CI ratings."

SIEMENS NX

SIEMENS NX (NX) is a software that allows simulations ranging from three-dimensional finite element modeling of conductive heat-transfer effects in building assemblies to modeling of airflow via computational fluid dynamics. The simulations are conducted on three-dimensional models that allow for the interaction of all the components within an assembly or transition. Transient simulations can be performed that include a time element to demonstrate the effects of weather on the heating and cooling of building components, or they can be performed as a static simulation showing the "worst-case" scenario. This type of analysis allows for a more accurate understanding of thermal performance and condensation risk due to the interactions of the various components in the assemblies.

This study is a continuation of a series of articles we have published previously. In the first publication, titled "Three-Dimensional Condensation Risk Analysis of Insulated Curtain Wall Spandrels,"[5] we evaluated how various methods of insulating spandrels, fire-rated joint methodologies, and application of various types of vapor retarders impact condensation risk at multiple locations on a curtain wall spandrels. In the second publication, titled "Three-Dimensional Effects on the Thermal Performance of Insulated Curtain Wall Spandrels,"[6] we explored the impact on the total product thermal performance (U-factor) of spandrel conditions based on several variables such as insulation type, insulation location, insulation thickness, fire-rated joint methodologies, and vapor retarder type and installation techniques. During those studies we utilized numerous correlation methods to calibrate the three-dimensional models to industry standard and proven two-dimensional modeling results. This current study relies on the previous correlation techniques.

Excerpts of Correlation Modeling from Previous Studies

The following are summaries of the correlation processes used in the previous two studies. The primary reason to include both two- and three-dimensional modeling is to continue to develop confidence in the results of the three-dimensional modeling procedures. Industry standard two-dimensional modeling has been tested and validated numerus times in the past for determining the total product U-factor and condensation risks of discrete vision units. Three-dimensional modeling is still a relatively new procedure in the glazing industry, and it is prudent to compare the modeling results from both methods.

The first step in this evaluation process included modeling the vertical sections of the curtain wall system with both types of software. This method omits the three-dimensional effects of the interfaces at the vertical and horizontal mullions. The results from the two- and three-dimensional programs were compared.

Figures 5 and 6 illustrate the results of both the two-dimensional and three-dimensional computer modeling procedures side by side for ease of comparison.

As indicated these figures, the resultant surface temperatures, produced by the two-dimensional and three-dimensional modeling procedures, are within one degree of each other. Based on the minimal deviation of corresponding temperatures between the models of the horizontal mullions, it was determined that the modeling procedures produce an acceptable level of correlation.

The resultant U-factors produced by the two-dimensional and three-dimensional modeling procedures in the second study are indicated in **figure 7** and are within a close margin of error. Based on the minimal deviation of thermal performance, the modeling procedures produce an acceptable level of correlation.

FIG. 5 Full section through four horizontal mullions, two IGUs, and one spandrel with foil-faced insulation (included from expert from previous publication).

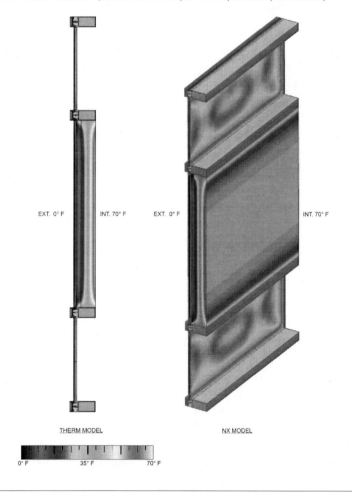

FIG. 6 Enlarged section details of sill and head horizontal mullion conditions at interface between vision unit and foil-faced spandrel (included from expert from previous publication).

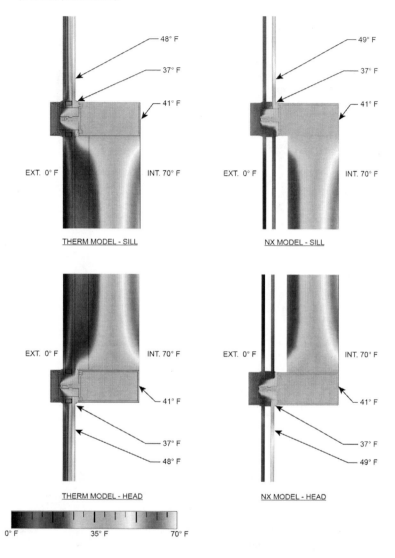

This continued analysis will further evaluate whether three-dimensional modeling is beneficial or necessary to predict the total product thermal performance and condensation risks or whether two-dimensional analysis is sufficient. To evaluate the risks and impact of the three-dimensional effects of a curtain wall parapet, both software methods are again utilized, and the results are compared.

FIG. 7 Full section through the baseline model in two dimensions and three dimensions (included from expert from previous publication).

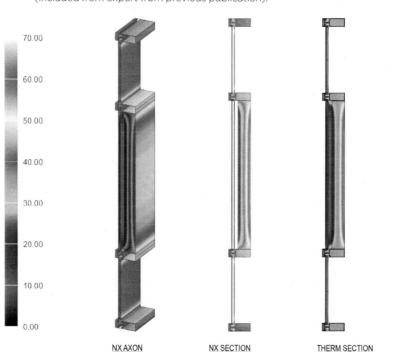

NX AXON

NX SECTION
SPANDREL U-FACTOR: 0.137
SPANDREL R-VALUE: 7.28

THERM SECTION
SPANDREL U-FACTOR: 0.135
SPANDREL R-VALUE: 7.41

SPECIMEN DESCRIPTION AND CONFIGURATIONS

Given the number of variables and due to the substantial amount of configurations, this evaluation fixed many of the variables that impact thermal performance to focus on factors that have a more significant contribution and that are related to the interface with the adjacent construction and have less to do with the curtain wall assembly components. Note that only select combinations of the various components indicated are included in this evaluation.

The following is a description of the base components that are fixed and remained unchanged throughout the evaluation:
- Curtain wall assembly:
  - Four-sided structurally glazed thermally broken unitized assembly
  - Spandrel glazing: 1-in. low-E-coated IGU with a stainless-steel edge spacer
  - Spandrel construction insulation/vapor barrier (where used):
  - 4-in. mineral wool
  - 22-gauge galvanized metal backpan vapor barrier

- Roof assembly:
  - 60-mil single-ply roof membrane
  - 6-in. polyisocyanurate insulation
  - 6-in. concrete roof deck
- Parapet assembly:
  - Parapet height: 3 ft. 6 in. tall
  - 6-in. cold-formed metal framing supported on top of the roof deck
  - Exterior gypsum sheathing on the backside of the metal framing
  - 3-in. polyisocyanurate insulation on the backside of the sheathing
  - Insulated wood coping box construction
  - Aluminum coping over parapet and curtain wall assemblies
- Fire-rated joint assembly
  - Aligned with roof deck
  - Mineral wool insulation with smoke seal

The following describes the variables that are modified throughout the evaluation:

- Intermediate horizontal curtain wall mullion:
  - No mullion
  - Fixed box mullion
  - Stack unitized mullion
- Location of horizontal mullion:
  - Aligned with top of roof deck/aligned with roof insulation
  - Aligned with bottom of roof deck
  - Placed below the roof deck
- Parapet insulation
  - Type 1
    - No insulation placed in the cold-formed metal framing wall
    - Continuous insulation on exterior side of parapet (roof side)
  - Type 2
    - Batt insulation placed in the cold-formed metal framing wall
    - Continuous insulation on exterior side of parapet (roof side)

Model Specimen Description and Parapet Configurations

The curtain wall specimens modeled in this evaluation are 9 ft. 6 in. tall and 7 ft. wide. Approximately 3 ft. 6 in. of the curtain wall extends into the parapet area. There is a typical horizontal mullion at the top and bottom of the curtain wall in all configurations, and in some configurations there is an intermediate horizontal mullion generally located near the roof line. There is an intermediate vertical mullion that extends from the interior space up into the parapet construction. **Figure 8** is an elevation of the typical curtain wall specimen with two section cuts indicated as AA and BB. Section cut AA (**fig. 9**) is through the typical area of the curtain wall, and

FIG. 8 Elevation of curtain wall specimen modeled in this evaluation.

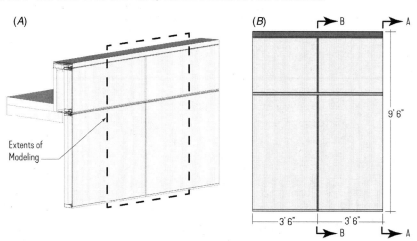

AA = SECTION THROUGH VERTICAL MULLION
BB = SECTION THROUGH MULLION CENTER OF UNIT

FIG. 9 Section cut AA—section through typical area of curtain wall specimen modeled in this evaluation.

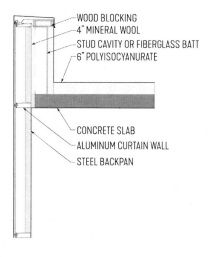

section BB (**fig. 10**) is cut though the center of the vertical mullion. **Figure 11** illustrates a typical specimen modeled in the three-dimensional software NX as viewed from the interior and exterior.

FIG. 10 Section cut BB—section through vertical mullion of curtain wall specimen modeled in this evaluation. Heat flows via conduction through the aluminum material and through the air cavity of the vertical mullions via convection as illustrated.

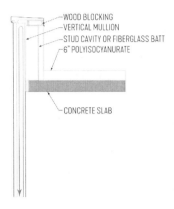

FIG. 11 Example of one of the curtain wall parapet specimens modeled in NX as viewed from the interior and exterior.

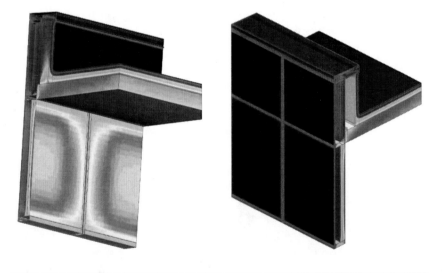

- **Parapet Configuration 1**: Stick-built curtain wall with no intermediate horizontal mullion located at the roof line
- **Parapet Configuration 2**: Stick-built curtain wall with an intermediate horizontal mullion located below the roof line
- **Parapet Configuration 3**: Stick-built curtain wall with an intermediate horizontal mullion below the roof line and with fiberglass batt insulation in the parapet framing

- **Parapet Configuration 4**: Stick-built curtain wall with an intermediate horizontal mullion aligned with the bottom of the roof line
- **Parapet Configuration 5**: Stick-built curtain wall with an intermediate horizontal mullion aligned with the top of the roof line
- **Parapet Configuration 6**: Unitized curtain wall with the horizontal stack mullion located below the roof line
- **Parapet Configuration 7**: Unitized curtain wall with the horizontal stack mullion located at the top of the roof line
- **Parapet Configuration 8**: Two independent stick-built curtain wall assemblies stacked on top of each other with a fiberglass thermal break material placed between two horizontal mullions generally aligned with the roof line

The environmental conditions used for analysis were as follows:
- Static interior and exterior temperatures.
- Exterior temperature: $-18°C$ ($0°F$).
- Exterior wind speed: 15mph.
- Exterior Film Coefficient[7]: 6 btu/h-ft^2-F
- Interior temperature: $21°C$ ($70°F$).
- Interior Film Coefficient[7]: 1.46 btu/h-ft^2-F
- Convective, Conductive, and Radiation heat and air flow are included in the modeling where applicable.

Temperature Color Gradient Legends

The two software programs used in this evaluation produce color legends that illustrate the output color as it relates to the temperature in the color gradient images. **Figure 12** includes examples the color legends as output from the two software programs.

THREE-DIMENSIONAL CONDENSATION RISK EVALUATION

As previously indicated, the eight parapet configurations are modeled in both THERM and NX. For each configuration, select temperatures are extracted from the models for comparison from both the section taken in the midspan of the horizontal mullion and from the section cut through the vertical mullion. Two sets of figures are provided for each configuration. Both sets include a conceptual image of the parapet, a two-dimensional color gradient output image from THERM, and a three-dimensional color gradient output image from NX. One set is where the section is cut through the midspan of the horizontal mullion, and the other set is where the section is cut through the vertical mullion. Temperatures taken from the NX model at the midspan of the horizontal mullion are not as significantly influenced by the three-dimensional effects of the vertical mullion and are typically comparable to those generated by the THERM model. Temperatures determined in the NX model where the section is cut through the vertical mullion are typically quite different from those in the THERM model, as the THERM model cannot represent the three-dimensional effects of the various components.

FIG. 12 (A) Temperature legend from THERM. (B) Temperature legend from NX.

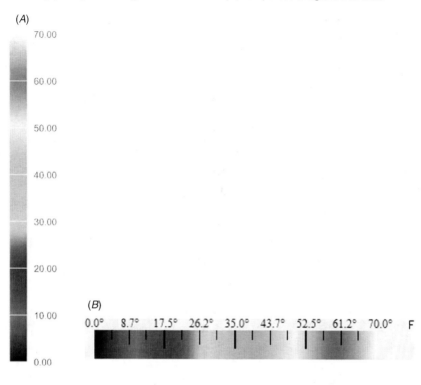

Configurations 1 through 5 incorporate stick-built curtain wall assemblies. With many stick-built curtain wall assemblies, the vertical mullions are typically not sealed and can allow interior air to enter the space in the vertical mullions and travel up into the parapet area. Depending on the relative humidity (RH) of the interior air, condensation can occur within the parapet area. Configurations 6 and 7 incorporate unitized assemblies that prevent the flow of interior air from entering the parapet by placing the horizontal stack mullion at the roof line. These configurations include an air seal to the bottom half of the stack mullion. The vertical mullions do not extend continuously into the parapet with this type of application. Configuration 8 completely divorces the parapet construction by utilizing two curtain wall assemblies separated by a fiberglass thermal break. This also prevents the vertical mullions from extending into the parapet.

The coldest temperatures that are exposed to the interior environment are extracted from the modeling output for evaluation. They are compared to each other and used to determine the maximum representative interior RH that the configuration can tolerate without the risk of condensation formation. Temperatures and RH levels are also compared at the top of the parapet construction.

Configuration 1

This configuration includes a stick-built curtain wall that does not incorporate an intermediate horizontal mullion at the roof line. As indicated by comparing the results of the section cut at midspan in **figure 13**, the cold point interior temperature on the THERM model is 61°F, which results in a maximum allowable RH of approximately 73%. The cold point temperature in the NX model of 50°F will only allow for a maximum RH of approximately 50%. If one were to rely on the THERM model, it would create a situation in which one might overestimate the capability of the system and assume that higher levels of interior RH can be tolerated.

The next step is to evaluate the models for which the vertical mullion extends into the parapet to determine whether there are colder temperatures present that would further define the allowable RH level. In **figure 14**, where the section is cut through the vertical mullion, the cold point temperature in the THERM model is only 15°F, which can tolerate a maximum interior RH of approximately 12%, whereas the NX model indicates a cold point temperature of 33°F, which can tolerate a maximum interior RH of approximately 25%. Due to the fact that THERM does not accurately account for the warming of the mullion due to the three-dimensional heat transfer, THERM in this case underestimates the performance of the assembly. The 25% RH level governs.

Based on the previously described issues with stick-built construction related to interior air bypassing the roof line through the vertical mullions, the top of the vertical mullion in the parapet needs to be evaluated to complete the analysis of this condition. The THERM model indicates a temperature of 1°F, and the NX model indicates 11°F. Although both are limited in the amount of RH that can be tolerated, the THERM modeling results may lead to a unnecessarily conservative requirement.

FIG. 13 (A) Section through curtain wall with no horizontal mullion at the roof line. (B) THERM model and results. (C) NX model and results.

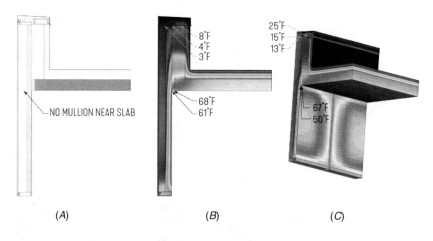

(A)　　　　　　　　　(B)　　　　　　　　　(C)

FIG. 14 (A) Section through curtain wall with no horizontal mullion at the roof line. Section is cut through the vertical mullion. (B) THERM model and results. (C) NX model and results.

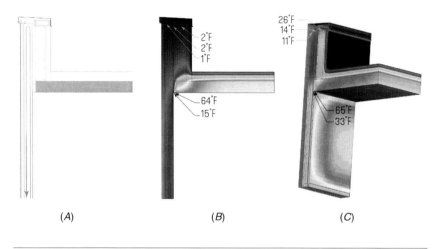

The maximum allowable RH for this configuration, based on the NX model, would be limited to 10%. However, if a method were implemented to ensure the vertical mullions are sealed to prevent migration into the parapet, then the maximum allowable RH would be 25%.

Configuration 2

This configuration includes a stick-built curtain wall with an intermediate horizontal mullion below the roof line. The results at the midspan in **figure 15** indicate a cold point temperature between 33°F and 35°F in the THERM and NX models, which tolerate a maximum RH between approximately 25% and 28%. In **figure 16** at the vertical mullion, the THERM model has a cold point temperature of 15°F, which can tolerate a maximum interior RH of approximately 12%, whereas the NX cold point temperature is 31°F, which can tolerate a maximum interior RH of approximately 24%. If the design set point of the interior RH was based on the THERM model, it could result in an unnecessarily low set point of 12%.

This configuration is similar to Configuration 1 and has the same issue with the unsealed vertical mullions. The THERM model indicates a temperature of 1°F, and the NX model indicates 9°F. Although both are again limited in the amount of RH that can be tolerated, the THERM modeling results is less accuracy, which would lead to a unnecessarily conservative requirement.

Again, the maximum allowable RH for this configuration, based on the NX model, would be limited to approximately 10%. However, if a method were implemented to ensure the vertical mullions are sealed to prevent migration into the parapet, then the maximum allowable RH would be 24%.

FIG. 15 (A) Section through curtain wall with horizontal mullion below the roof line. (B) THERM model and results. (C) NX model and results.

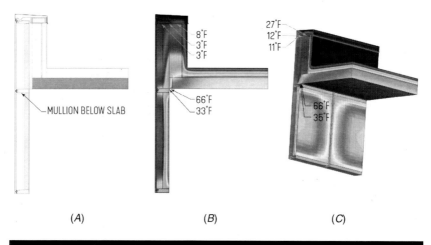

FIG. 16 (A) Section through curtain wall with horizontal mullion below the roof line. Section is cut through the vertical mullion. (B) THERM model and results. (C) NX model and results.

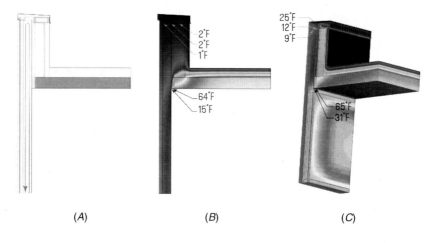

Configuration 3

This configuration incorporates a stick-built curtain wall assembly with the intermediate horizontal mullion below the roof line and with fiberglass batt insulation in the parapet framing. The only difference between Configuration 3 and Configuration 2 is that this specimen includes fiberglass batt insulation in the parapet framing. The purpose of including this variable is to evaluate the benefit of the added insulation. When comparing the results in **figures 15** and **16** to **figures 17** and **18**, it is clear that there is limited value in adding the insulation, as the temperatures did not vary in any significant manner. This exercise was performed on many of the

FIG. 17 (A) Section through curtain wall with horizontal mullion below the roof line and batt insulation installed in the parapet wall. (B) THERM model and results. (C) NX model and results.

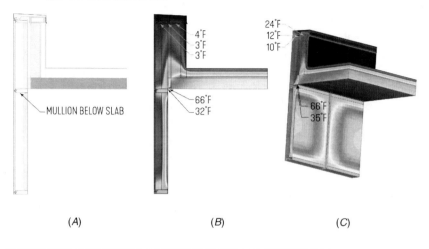

FIG. 18 (A) Section through curtain wall with horizontal mullion below the roof line and batt insulation installed in the parapet wall. Section is cut through the vertical mullion. (B) THERM model and results. (C) NX model and results.

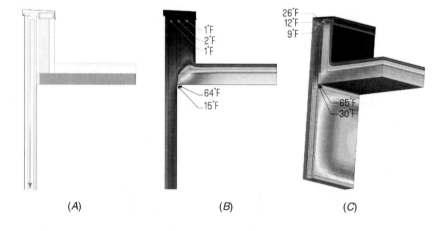

other configurations in this study but are not included because they resulted in a similar minor impact.

Note that the NX model indicates 9°F at the top of the vertical mullion, which again limits the maximum allowable RH for this configuration to approximately 10%. Again, if a method were implemented to ensure the vertical mullions are sealed to prevent migration into the parapet, then the maximum allowable RH would be 24%.

Configuration 4

This configuration incorporates a stick-built curtain wall with horizontal mullion aligned with the bottom of the roof line. This configuration is also similar to Configuration 2, but with the intermediate horizontal mullion raised upward slightly to align with the bottom of the roof line. The purpose of this evaluation is to determine whether the location of the horizontal mullion relative to the fire-rated joint would have an impact on the allowable RH. As indicated in the results displayed in **figures 19** and **20**, the shift in the horizontal mullion location had a minimal impact

FIG. 19 (*A*) Section through curtain wall with horizontal mullion aligned with the bottom of the roof line. (*B*) THERM model and results. (*C*) NX model and results.

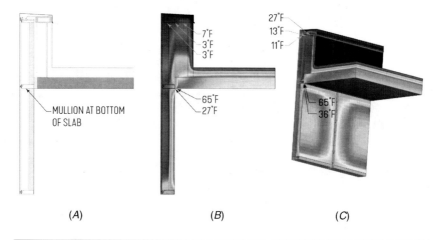

FIG. 20 (*A*) Section through curtain wall with horizontal mullion aligned with the bottom of the roof line. Section is cut through the vertical mullion. (*B*) THERM model and results. (*C*) NX model and results.

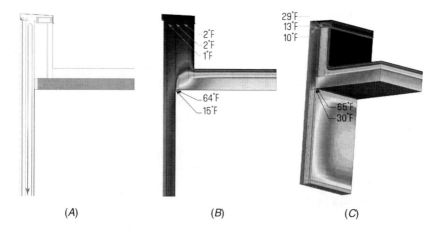

on the overall condition, as the cold point is still at the vertical mullion where it interfaces with the fire-rated joint and did not change significantly.

Similar to Configurations 1–3 the NX model indicates 10°F at the top of the vertical mullion, which limits the maximum allowable RH for this configuration to approximately 10%. Again, if a method were implemented to ensure the vertical mullions are sealed to prevent migration into the parapet, then the maximum allowable RH would be 24%.

Configuration 5

This configuration incorporates a stick-built curtain wall with horizontal mullion aligned with the top of the roof line. This configuration is also similar to Configuration 2 and Configuration 4, but with the intermediate horizontal mullion raised further upward to align with the top of the roof line. The purpose of this evaluation is to determine whether the location of the horizontal mullion relative to the fire-rated joint would have an impact on the allowable RH. As indicated in the results displayed in **figures 21** and **22**, the shift in the horizontal mullion location had a minimal impact on the overall condition, as the cold point is still at the vertical mullion where it interfaces with the fire-rated joint and did not change significantly.

Similar to Configurations 1–4 the NX model indicates 9°F at the top of the vertical mullion, which limits the maximum allowable RH for this configuration to approximately 10%. Again, if a method were implemented to ensure the vertical mullions are sealed to prevent migration into the parapet, then the maximum allowable RH would be approximately 24%.

Configuration 6

This configuration incorporates a unitized curtain wall with the horizontal stack mullion located below the roof line. As previously indicated, this configuration

FIG. 21 (A) Section through curtain wall with horizontal mullion aligned with the top of the roof line. (B) THERM model and results. (C) NX model and results.

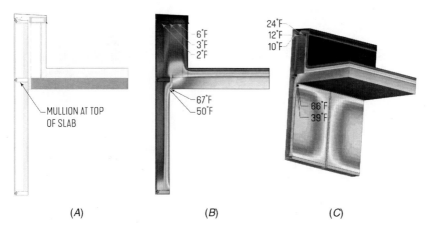

FIG. 22 (A) Section through curtain wall with horizontal mullion aligned with the top of the roof line. Section is cut through the vertical mullion. (B) THERM model and results. (C) NX model and results.

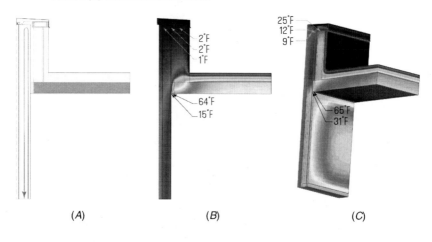

includes the horizontal stack joint that does not provide a path for interior air to flow through the curtain wall assembly up into the parapet. Similar to the other configurations, the cold point temperature that will govern the interior RH is near the vertical mullion where it interfaces with the fire-rated joint within the horizontal stack mullion.

As indicated in **figures 23** and **24**, the cold point temperature is 26°F and is taken from the NX model, which results in a maximum allowable RH of approximately 20%. The cold point is located on the underside of the bottom half of the

FIG. 23 (A) Section through unitized curtain wall with horizontal stack mullion at the bottom of the roof line. (B) THERM model and results. (C) NX model and results.

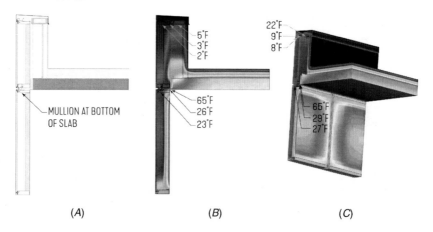

FIG. 24 (A) Section through unitized curtain wall with horizontal stack mullion at the bottom of the roof line. Section is cut through the vertical mullion. (B) THERM model and results. (C) NX model and results.

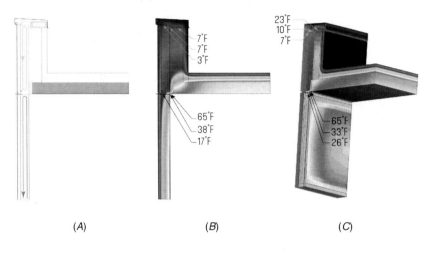

(A)   (B)   (C)

horizontal stack mullion. This is a critical location because interior air can enter the vertical mullions at the next horizontal stack joint below. Note that because the stack joint does not allow air to enter the parapet, temperatures above the stack joint do not need to be evaluated.

Configuration 7

This configuration incorporates a unitized curtain wall with the horizontal stack mullion located at the top of the roof line. This configuration is similar to Configuration 6, but with the horizontal stack mullion raised upward to align with the top of the roof line. The purpose of this evaluation is to determine whether the location of the horizontal mullion relative to the fire-rated joint would have an impact on the allowable RH. As indicated in the results displayed in **figures 25** and **26**, the shift in the horizontal mullion location had an impact on the overall condition, and the cold point that will govern the interior RH is near the vertical mullion where it interfaces with the fire-rated joint within the horizontal stack mullion.

The cold point temperature is 20°F and is taken from the NX model, which results in a maximum allowable RH of approximately 15%. The cold point is located on the underside of the bottom half of the horizontal stack mullion. This is a critical location because interior air can enter the vertical mullions at the next horizontal stack joint below. Note that because the stack joint does not allow air to enter the parapet, temperatures above the stack joint do not need to be evaluated.

Configuration 8

This configuration includes two independent stick-built curtain wall assemblies stacked on top of each other with a fiberglass thermal break material placed

FIG. 25 (A) Section through unitized curtain wall with horizontal stack mullion at the top of the roof line. (B) THERM model and results. (C) NX model and results.

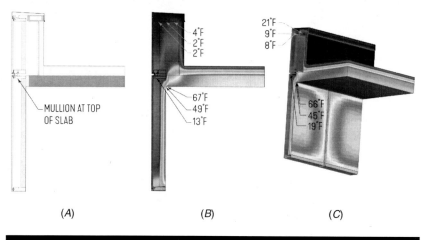

FIG. 26 (A) Section through unitized curtain wall with horizontal stack mullion at the top of the roof line. Section is cut through the vertical mullion. (B) THERM model and results. (C) NX model and results.

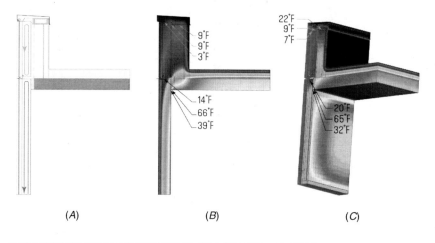

between two horizontal mullions generally aligned with the roof line. Similar to the configurations that included the unitized curtain wall, this configuration is sealed at the thermal break location and does not provide a path for interior air to flow up into the parapet through the curtain wall. Similar to the other configurations the cold point temperature that will govern the interior RH is near the vertical mullion where it interfaces with the fire-rated joint within the at the bottom side of the fiberglass thermal break.

As indicated in **figures 27** and **28**, the cold point temperature is 25°F and is taken from the NX model, which results in a maximum allowable RH of

FIG. 27 (A) Section through the two stick built curtain walls with the two horizontal mullions and thermal break aligned with the roof line. (B) THERM model and results. (C): NX model and results.

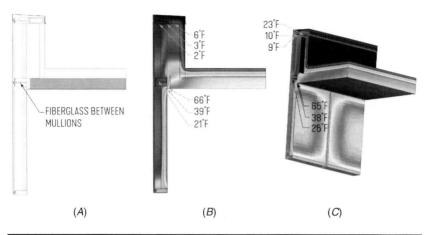

FIG. 28 (A) Section through the two stick built curtain walls with the two horizontal mullions and thermal break aligned with the roof line. Section is cut through the vertical mullion. (B) THERM model and results. (C) NX model and results.

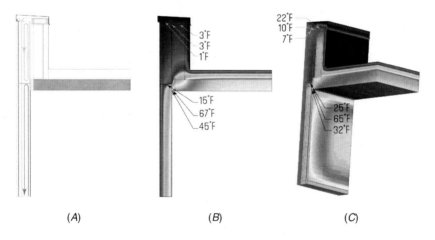

approximately 20%. Note that because the air seal is presumed to be located at the fiberglass thermal break, air is not able to enter the parapet and the temperatures above do not need to be evaluated.

THREE-DIMENSIONAL THERMAL PERFORMANCE EVALUTION
Due to the inherent thermal bridging of the curtain wall that extends into the parapet area, it is reasonable to presume that this condition would lead to reduced

thermal performance of the overall interface between the roof and wall. The NX output was used to determine a total product U-factor for the portion of the curtain wall assembly exposed to the interior, which includes the heat flow out through the parapet. Additionally, the entire composite U-factor that includes the roof and the curtain wall was determined. **Figure 29** illustrates the extent of the areas included in the curtain wall U-factor and the composite U-factor. This information was determined for all parapet configurations to better understand the impact of each modification. The results of the curtain wall thermal performance and composite performance are summarized in **table 1**. As indicated by the results included the summary, each configuration had a similar impact on the overall performance.

FIG. 29 Extent of curtain wall U-factor and the composite U-factor.

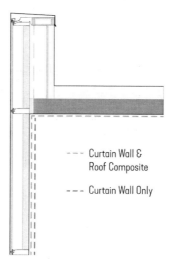

TABLE 1 Summary of total product thermal performance

Parapet Configuration	Curtain Wall/Roof Composite		Curtain Wall Only	
	U-Factor	R-Value	U-Factor	R-Value
1	0.188	5.32	0.259	3.87
2	0.210	4.76	0.288	3.47
3	0.204	4.91	0.288	3.47
4	0.204	4.89	0.277	3.61
5	0.198	5.05	0.269	3.71
6	0.208	4.82	0.284	3.53
7	0.193	5.19	0.252	3.97
8	0.196	5.09	0.288	3.47

Summary and Conclusions

It is important to note that the results of this evaluation are generally applicable only to specific set of environmental conditions, material selections, assembly types, and configurations used in the evaluation. Although some of the results obtained from the specific configurations may be able to be used to inform similar situations, care must be taken when inferring the results of this evaluation to other situations if specific modeling is not performed. The following are some key observations made when evaluating the configurations used in this evaluation.

The location of the intermediate horizontal mullion in both stick-built and unitized assemblies has moderate impact on the condensation resistance at the midspan of the horizontal mullion. The best placement is to locate the mullion higher in the assembly and as indicated in Configuration 1. Eliminating the mullion altogether provides the best performance at the midspan. This is due to the fact that the horizontal mullion is also a thermal bridge from the interior to the parapet. Although this has been proven for the models conducted in this study, this may be different when working with a vision unit below the roof line. However, in all of the configurations, the vertical mullion has the coldest point, which did not change significantly when modifying the location of the intermediate horizontal mullion and ultimately governs the level of interior RH that can be tolerated.

As previously indicated, in many stick-built curtain wall assemblies, the vertical mullion ends are typically not sealed and can allow interior air to enter the space in the vertical mullions and travel up into the parapet area. This will typically be the governing factor for the level of interior RH the system can tolerate without the risk of condensation. Note that in cold climates this could lead to a reduced ability to humidify during winter conditions and can be a condition that is easily overlooked. There may be a means with certain curtain wall system to provide an airtight seal at the vertical mullions. However, accomplishing the air sealing can be complex and challenging to achieve and verify.

Many unitized curtain systems, represented in Configurations 6 and 7, can introduce a horizontal stack mullion at the roof line that can prevent the flow of interior air from entering the parapet. The vertical mullions do not extend into the parapet with this type of application. Implementing a method similar to Configuration 8 that completely divorces the parapet construction will also prevent the vertical mullions from extending into the parapet. Both require providing an airtight seal at the roof line and performed similarly.

The addition of insulation within the parapet stud cavity has a negligible effect on the performance of the fenestration. This is likely due to the reduction of thermal performance of the batt insulation from the bridging of the studs. Not only is there thermal bridging of the stud from the interior (back of spandrel) to the exterior (roof side), but there is also thermal bridging vertically (from roof deck to the coping). The configurations in this study included continuous roof insulation on the roof side of the framing. This tempers the potential benefit from the added batt insulation.

This study compared a limited number of curtain wall parapet conditions with a select number of variables. The results indicate when curtain walls are incorporated into parapet constructions, the level of interior relative humidity needs to be considered and likely constrained to lower levels to prevent the risk of condensation. Blocking the airflow into the parapet is a key factor in the ability to accommodate increased levels of interior humidity. There are other design and construction methods that may allow for increased levels of interior humidity beyond what was determined in this study.

There is a significant amount of continued evaluation that still needs to be performed to understand the issue of cantilevered curtain wall components better. Future studies may include other conditions such as:
- Varying parapet height
- Elimination of parapet/suspension of separate curtain wall above
- Varying parapet insulation thickness
- Use of vision units below roof line
- Use of vision units above the roof line
- Different spandrel/shadow box options
- Unitized mullion types (single vs. double chicken head at horizontal stack mullion)
- Thickness and materiality of horizontal curtain wall system isolator/thermal break

Although there is still significant work that can be performed, this evaluation continued to support the use of three-dimensional thermal modeling. As demonstrated by the examples used in this evaluation, there are conditions that cannot be truly understood in a two-dimensional modeling process. If relying solely on a two-dimensional modeling process, it is possible to become overly conservative in selecting the threshold of interior RH that can be tolerated without causing condensation to occur, or there may be locations within the assemblies that can be overlooked that are colder than anticipated and might lead to an overly optimistic threshold. Either situation is plausible, and the use of three-dimensional modeling can provide additional insight when selecting an acceptable interior RH level. Although correlation of the two different modeling techniques was not a specific topic of this evaluation, we continue to monitor the results of the two methods. General correlation was periodically reviewed at locations where the three-dimensional effects of the assemblies should have a limited impact on the results. When reviewing these types of locations in the eight configurations, we continued to observe reasonable correlation. However, there is still limited data available in the industry comparing the three-dimensional modeling to physical testing results. This effort is still an opportunity for our industry to further validate three-dimensional modeling as a viable method to reduce the need for future physical testing of project-specific conditions.

References

1. *Voluntary Test Method for Thermal Transmittance and Condensation Resistance of Windows, Doors and Glazed Window Sections*, AAMA 1503.1 (Schaumburg, IL: American Architectural Manufacturers Association, 1988).
2. *Procedure for Determining Fenestration Product Condensation Index Ratings*, ANSI/NFRC 500 (Greenbelt, MD: National Fenestration Rating Council, 2013).
3. *Standard Test Method for Thermal Performance of Building Materials and Envelope Assemblies by Means of a Hot Box Apparatus*, ASTM C1363-19 (West Conshohocken, PA: ASTM International, approved September 1, 2019), https://doi.org/10.1520/C1363-19
4. *Procedure for Determining Fenestration Product U-Factors*, ANSI/NFRC 100 (Greenbelt, MD: National Fenestration Rating Council, 2013).
5. R. Asava and A. Dunlap, "Three-Dimensional Condensation Risk Analysis of Insulated Curtain Wall Spandrels," in *Advances in Hygrothermal Performance of Building Envelopes: Materials, Systems and Simulations*," ed. P. Mukhopadhyaya and D. Fisler (West Conshohocken, PA: ASTM International, 2017).
6. R. Asava, K. Gross, and A. Dunlap, "Three-Dimensional Effects on the Thermal Performance of Insulated Curtain Wall Spandrels" (paper presentation, fifth BEST Conference Building Enclosure Science and Technology, Philadelphia, PA, April 15–18, 2018).
7. ASHRAE, *ASHRAE Handbook: Fundamentals* (Atlanta, GA: Author, 2017).

STP 1635, 2022 / available online at www.astm.org / doi: 10.1520/STP163520200120

Andrea Wagner Watts[1] and William Ranson[2]

Managing Interfaces in Complex Systems

Citation

A. W. Watts and W. Ranson, "Managing Interfaces in Complex Systems," in *Building Science and the Physics of Building Enclosure Performance: 2nd Volume*, ed. D. J. Lemieux and J. Keegan (West Conshohocken, PA: ASTM International, 2022), 174–194. http://doi.org/10.1520/STP163520200120[3]

ABSTRACT

Continuity of the air and water-resistant barrier (AWB) is critically important in high-performing buildings. Achieving continuity of these barriers at interfaces of different material systems is often a challenge. It becomes more challenging when the different barriers are installed in a nontraditional order within the building envelope or when penetrations are added after the AWB has been installed. This paper looks at the design and installation of tested, durable transitions between the AWB and other systems, including considerations for the window interface when the AWB is installed outboard of the continuous insulation, when the insulation is installed behind the exterior sheathing and AWB, and when the continuous insulation is also the AWB in commercial construction. The paper also examines the effects of the AWB being in or out of plane with the windows and the effects of the sequencing of window installation, specifically what happens when the window is installed either before or after the AWB. Options to reduce the number of fastener penetrations to the AWB before cladding installation are also shown. All discussed interface details are based on results from advanced system testing using a sequence of multiple ASTM test methods for air and water penetration before and after structural and thermal conditioning. In addition to interfaces in place prior to the installation of the AWB, penetrations are often installed after the AWB. One of the most frequent

---

Manuscript received January 4, 2021; accepted for publication August 6, 2021.
[1]DuPont Performance Building Solutions, 174 Main St. #107, East Aurora, NY 14052, USA http://orcid.org/0000-0002-5399-4761
[2]DuPont Performance Building Solutions, Donnelly 106, 5401 Jefferson Davis Hwy., Richmond, VA 23234, USA http://orcid.org/0000-0002-8154-5189
[3]ASTM Second Symposium on *Building Science and the Physics of Building Enclosure Performance* on April 24–25, 2022, and June 12, 2022 in Seattle, WA, USA.

Copyright © 2022 by ASTM International, 100 Barr Harbor Drive, PO Box C700, West Conshohocken, PA 19428-2959.

ASTM International is not responsible, as a body, for the statements and opinions expressed in this paper. ASTM International does not endorse any products represented in this paper.

penetrations is for cladding attachments such as girt systems and when continuous insulation is installed. This paper presents an overview of results from testing conducted to determine how best to attach girt systems to continuous insulation that is also performing as the AWB along with an introduction to new innovations that ensure long-term water tightness when installing continuous insulation over an AWB.

Keywords
air barrier, water-resistant barrier, interfaces, window interface, testing, continuity, penetrations, continuous insulation

Introduction

The building envelope can be defined as the area separating conditioned and unconditioned space or everything separating the interior of a building from the outdoor environment.[1] The use of defined control layers for preventing air, water, thermal, and vapor infiltration through the building envelope became industry standard more than a decade ago. The control layers are designed to minimize the movement of air and liquid water, slow the movement of water vapor, and reduce heat transfer through the building envelope—both into the building and out.

The "perfect wall" as defined by Joseph Lstiburek shows all four independent layers on the exterior of the building envelope.[2] This may be the easiest approach to define, but there are many ways to meet the same needs of a high-performance building envelope by combining the control layers into fewer products or even putting them in a different order within the wall assembly. Many air barriers on the market are designed to go on the exterior of a building and can also be used as a water-resistant barrier when detailed properly. These materials have been designed and tested to meet the requirements of both control layers and are referred to as air and water-resistant barriers (AWBs). AWBs can also be a vapor retarder. If one of these "3-in-1" materials is used on the exterior of the building, it is critically important to not have an additional vapor control layer on the interior of the wall assembly such as is prescribed in some building codes. The use of two vapor retarders in a wall assembly can cause mold and early degradation of the wall assembly if not designed and installed properly.[3] This is especially true for steel-framed construction.

There are some products on the market, such as insulating foam sheathing and medium-density spray polyurethane foam, that can serve as all four control layers—providing air, water, thermal, and water vapor protection for the building when installed according to the manufacturer recommendations. For foil-faced polyisocyanurate (ISO) insulation, the aluminum facer provides the air, water, and vapor control layers. The seams between the insulation boards must be sealed to maintain air and water tightness of the system, and specific fasteners must be used to secure the boards to the structure. When there is a reliance on a single material to perform the functions of all four control layers, the need for robust testing of the system is critical. This includes air infiltration assembly testing that is required for air barriers as well as additional testing for use of the system as a water-resistant barrier. ICC-ES Acceptance Criteria 71 (AC71) includes the requirements for how to test to confirm the seam treatment within ISO insulation systems when used as a water-resistant barrier.[4]

In addition to combining multiple control layers into one material, the order in which the layers are installed on the building can be changed on the basis of the building type, climate zone, and compatibility of materials. Although the most common building practice is to install the AWB to sheathing behind the continuous insulation (CI), it is also possible to remove the sheathing, install the CI directly to the studs, and then install an AWB on the exterior of the CI. When changing the order of the control layers, it is especially important to pay attention to where the thermal and vapor control layers will be placed. The thermal control layer can also be a point of vapor control depending on the type of insulation and how the assembly is detailed. Additionally, owing to the reduction in thermal bridging, the insulation will have a higher efficacy when put external to the studs.[5] If the vapor control layer is separate from the thermal control layer, the thermal and vapor control layers need to be properly designed to reduce the potential for mold growth within the wall assembly. The vapor control layer can be on the interior or exterior of the wall assembly: it is most often put on the warm-in-winter side of the thermal control layer or the warm-in-winter side of the wall assembly depending on the climate zone. Special analysis may be required for buildings with a higher interior relative humidity to determine where the vapor control layer is best located. Design techniques that include the use of hygrothermal modeling can help to make these decisions.

No matter the location of the control layers, it is important to ensure proper detailing and continuity of the air, water, and thermal control layers at interfaces within the building envelope. Discontinuities at the roof, foundation, or window transitions can negate the hard work done to install the control layers in the field of the opaque wall. Additionally, any time that the control layers are penetrated—by piping, cladding attachments, ductwork, and so on—it is especially critical to ensure that the air and water control layers have not been compromised. This can be done prior to construction through lab assembly testing as discussed in this paper or by using whole building testing after construction is complete. Although whole building air tightness testing is becoming a code requirement in certain jurisdictions, the location of the air leakage can be difficult to find and repair.

Assembly Testing

The code requirements for air barrier assemblies within the International Energy Conservation Code (IECC) reference ASTM E2357, *Standard Test Method for Determining Air Leakage Rate of Air Barrier Assemblies*.[6] This test method specifies exact designs of the wall assembly, including penetrations, a window opening, and brick ties. The assembly is tested for air leakage per ASTM E283, *Standard Test Method for Determining Rate of Air Leakage through Exterior Windows, Skylights, Curtain Walls,* and Doors under Specified Pressure Differences across the Specimen,[7] both before and after wind pressure or structural conditioning to ensure the system has less than a code-mandated maximum air leakage rate. The current IECC does not include a requirement for additional types of penetrations such as the large fasteners usually required to attach cladding or insulation after the air barrier is installed.

There are additional requirements for water-resistant barriers (WRBs) found in the International Building Code (IBC). Although WRBs are required by code and a clause for testing new materials for use as a WRB is included, there is variability in what testing is required to be completed. The ICC-ES code acceptance criteria for WRBs (which are not necessarily a code requirement depending on the material and jurisdiction) often require the ASTM E2357 penetrated assembly (Wall 2) to be tested for water infiltration per ASTM E331, *Standard Test Method for Water Penetration of Exterior Windows, Skylights, Doors, and Curtain Walls by Uniform Static Air Pressure Difference*.[8] The exact pressures and length of the testing, as well as the type of wall assembly needed to be tested, vary by material. As an example, the ICC-ES criteria for fluid-applied, self-adhered, and mechanically fastened WRBs requires the materials to be tested for 15 min at 6.24 psf (300 Pa). Per ICC-ES AC 71, exterior foam sheathing must be tested for 2 h at the same pressure of 6.24 psf (300 Pa).

Full system testing of multilayer systems is rarely part of the development testing of air barrier systems, so determining the best way to detail interfaces of complex building systems is often left to the contractor or design team. This is starting to change. There are two general strategies for testing of materials as assemblies: testing to code minimums and testing past code minimum to specific design criteria. In some instances, assemblies are tested past code minimum to try to predict performance in specific applications or projects. Assemblies from standardized test methods can be modified to ensure that all necessary interfaces are included in the assembly and to include more control layers in the testing to ensure that all the different materials can maintain their properties when constructed. One example of this is to install the AWB on exterior sheathing and then install CI over the AWB and finally add cladding attachments such as hat channels or brick ties through all the layers back to the structure. This assembly can be tested for air and water penetration to ensure there is no uncontrolled leakage.

Wall assemblies discussed in this research have been tested to a full testing protocol like that used for evaluation of some AWBs with additional testing for better assurance of long-term robustness. The overall series of tests is similar to those found in AAMA 504, *Voluntary Laboratory Test Method to Qualify Vertical Fenestration Installation Procedures*, for testing window installation assemblies.[9] The same wall assembly is tested for both air and water infiltration before and after different types of conditioning such as structural loading and thermal cycling. This helps determine whether either of these stressors causes the assembly to become more porous to air, or more importantly for this study, water. It is not uncommon for wall assemblies that pass the current code limits for air infiltration to have issues with water leakage when tested to this protocol. There are slight differences in the test protocol and pressures for residential and commercial construction types based on the performance needs of the completed building.

The general test protocol used to evaluate the commercial wall assemblies addressed in this paper is as follows:
- Air infiltration testing per ASTM E283 at up to 300 Pa (6.24 psf) pressure differential with special focus on the code compliance level of 75 Pa (1.57 psf).
- Water infiltration testing per ASTM E331 under a negative pressure ramping up to 300 Pa (6.24 psf) held for 2 h before continuing at different levels up to 720 Pa (15 psf), being held at each of the intermittent and maximum pressures for 15 min each.
- Wind pressure conditioning per ASTM E2357. This portion of the testing is often referred to as structural loading, which includes 1-h sustain loading at 600 Pa, 2,000 cycles of cyclic loading at 800 Pa, and 3-s gust loads at 1,200 Pa. All testing is completed at positive and negative pressures.
- Repeat air filtration testing per ASTM E283.
- Repeat water infiltration testing per ASTM E331.
- Thermal cycling per ASTM E2264-05(2013), *Standard Practice for Determining the Effects of Temperature Cycling on Fenestration Products*, Method A, Level 2.[10]
- Repeat air filtration testing per ASTM E283.
- Repeat water infiltration testing per ASTM E331.

The general test protocol used to evaluate the residential, wood-framed wall assemblies addressed and recommended in this paper is similar to the commercial protocol above with a couple key differences:
- Water infiltration testing per ASTM E331 is tested to a maximum of 300 Pa (6.24 psf) held for 15 min.
- Wind pressure conditioning is completed per ASTM E1677 by holding at 515 Pa (10.8 psf) for 1 hour both infiltration and exfiltration.[11]

The pass/fail criteria for this testing follows the code requirements and those in the respective standards: an air leakage rate less than 0.2 L/s·m^2 (0.04 cfm/ft.^{2}) at a pressure differential of 75 Pa (1.57 psf), no water observed on the interior side of the wall assembly during any of the water penetration testing, and no physical damage observed during the structural loading.

Design and Installation of Nontraditional Wall Assemblies

AIR AND WATER-RESISTANT BARRIERS INSTALLED OVER CONTINUOUS INSULATION FOR COMMERCIAL AND RESIDENTIAL CONSTRUCTION

Installing the AWB over the CI instead of the traditional method of installing the CI over the AWB can be useful in some buildings (**fig. 1**). First, when foam sheathing is used as the CI, it can be installed directly against the framing or stud walls.

FIG. 1 AWB over foam sheathing versus AWB under CI in a commercial wall assembly.

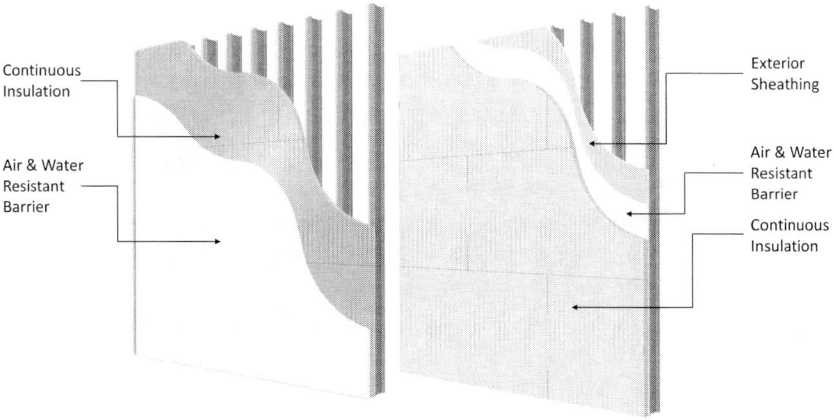

Continuous Insulation

Air & Water Resistant Barrier

Exterior Sheathing

Air & Water Resistant Barrier

Continuous Insulation

This eliminates the need for exterior sheathing when it is not needed for structural support or fire requirements. Having the AWB on the exterior of the assembly allows for easier verification of proper fastener treatment of the insulation fasteners and any AWB fasteners if mechanically fastened AWBs are used. Another advantage of this assembly is that it allows for the detailing of penetrations (such as cladding fasteners) at the point at which they breach the AWB layer. This detailing is often required to maintain air and water tightness following the installation of girts and other types of cladding attachment, especially for lightweight cladding. In a traditional assembly, it is challenging to detail any later penetrations at the interface of the AWB because it is difficult to reach back through the CI without damaging it. Instead, the penetration is often treated at the face of the CI, which does not allow for continuity of the air and water control layers in that assembly.

We have performed multiple water infiltration assembly tests per ASTM E331 that have shown that treating penetrations in this manner is inconsistent at best, even at low pressures (below 6.24 psf). The results of this testing, although not published in themselves, are the foundation for the types of fastener sealing solutions recommended by AWB manufacturers.[12] There are also fastener manufacturers that have developed fastening solutions that can be installed through the CI and seal at the point of penetration of both the CI and the AWB.

There are some important items to consider when selecting this type of wall assembly. The first consideration is the service temperature limits of the AWB. When the AWB is no longer protected by the CI, it will likely be exposed to more extreme temperatures than are typical for a traditional wall assembly in the given climate zone. Second, analysis for vapor control must be completed for the wall assembly in the climate zone. If nonpermeable insulation is selected, the vapor barrier will be in a similar position to a traditional wall assembly. However, that is not true for permeable insulation. When fluid-applied AWBs are used over CI, the elongation and recovery of the AWB and the flashing or sealant used to seal the joints of the CI must be considered. Continuous insulation, just like all other materials, will expand and contract under changing temperature conditions. For example, extruded polystyrene has a coefficient of linear thermal expansion of 3.5×10^{-5}, which means a standard 4' × 8' board can expand or contract upwards of 1/8" under standard service conditions of the material. The required elongation of the material used to seal the seams between the CI boards must be able to take this movement.

Sufficient CI must be used such that the dew point within the wall assembly falls to the interior of the CI to prevent condensation within the wall cavity. This can be demonstrated with a dew-point analysis of a steel stud assembly as shown in **figure 2**. In cold climates, when the insulation is split between the stud cavity and CI, the dew point can fall within the study cavity, which results in the risk of condensation on the interior of the exterior gypsum (**fig. 2**, Assembly A). However, when all of the insulation is put outside the stud cavity, the risk of condensation is eliminated because the dew point is not completely outside the AWB (**fig. 2**, Assembly B). Dew-point analysis is linear and assumes steady-state temperature conditions, so a more detailed hygrothermal analysis may be required for specific projects to ensure moisture levels within the different components of the wall assembly remain at acceptable levels throughout the life of the building.

Testing has shown that other specific details are slightly different when installing the AWB over the continuous insulation. When treating rough openings in this type of assembly, it is important that the AWB wrap over the edges of the insulation and back into the window opening. This may require wider pieces of

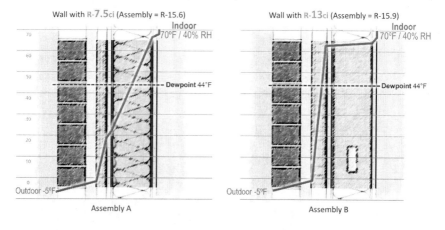

FIG. 2 Dew point analysis showing impact on dew point location when continuous insulation is moved outside the stud cavity.

self-adhered flashing than is traditionally used, or it may require the use of a wood bump out. If fluid-applied flashing is used to treat the window opening, it is critical to evaluate the compatibility of the flashing with the CI so premature degradation of the CI does not occur. Similar consideration must be taken when detailing the roof and foundation interfaces. The roof-to-wall interface is very similar to a traditional detail with the AWB wrapping over the top of the CI before tying into the roof air and water control layers. The foundation interface may take additional detailing because the AWB may need to wrap under the CI to tie into the foundation waterproofing. If this occurs, the workmanship of applying and rolling any self-adhered materials onto the underside of the wall barriers is critical to the success of the transition. The order of installation of the foundation protection in comparison to the wall AWB and CI has a strong influence on how this detail is accomplished.

Care must be taken when installing the AWB over CI as well. When fluid-applied AWBs are used over CI, it is important to confirm adhesion and compatibility of the two materials. For example, extruded or expanded polystyrene insulation may degrade when specific fluid-applied coatings are installed over them and then exposed to ultraviolet light. When applying a fluid-applied AWB over CI, it is important to make sure that there are no pinholes or cracks at any of the CI fasteners. This may require pretreatment of the fasteners in a way similar to treating screw holes of exterior gypsum prior to the installation of the AWB. The use of self-adhered AWBs over foam sheathing insulation may be a challenge to install properly depending on how easily it is to back-roll the material around the CI fasteners. For these reasons, mechanically fastened AWBs may be the easiest solution for these types of assemblies.

INVERTED WALL ASSEMBLIES FOR COMMERCIAL CONSTRUCTION

Another wall assembly solution for installing the AWB outboard of CI in steel stud construction is to build an inverted wall (fig. 3). In this assembly, insulating sheathing insulation is installed directly to the framing. Exterior sheathing is then installed outboard of the CI followed by the AWB. This assembly contains all the traditional control layers as well as the exterior sheathing found in many wall assemblies. This assembly also allows for the AWB to be installed and detailed much the same way as outlined by general manufacturer installation instructions. Like the previous assembly, it also allows for any later penetrations into the wall assembly to be detailed at the AWB layer. This greatly simplifies how to detail cladding attachments that are installed over the assembly as any fasteners do not have to be sealed through the insulation. An inverted wall assembly also allows for the insulating sheathing to be protected by exterior gypsum. This does two things: (1) it removes any potential issues of compatibility between the AWB and the CI; and (2) it can allow a larger variety of assembly options that meet NFPA 285 requirements when using insulating foam sheathing.

There are special considerations needed for installing this wall assembly. When the exterior sheathing is installed over the CI, it is important to make sure that the exterior sheathing fasteners are long enough to tie back to the structure. The same

FIG. 3 Inverted wall assembly showing extruded polystyrene against the studs with AWB over exterior gypsum.

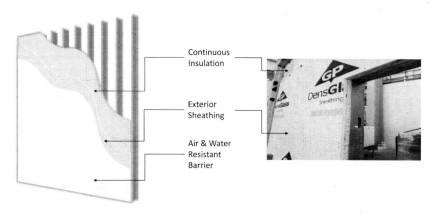

is true for cladding attachment fasteners. As is true anytime cladding attachments are secured over CI, the compressive strength of the insulation must be able to tolerate the load. Chapter 26 of the International Building Code contains tables that outline when additional furring is required to secure cladding over foam sheathing based on the wall structure, fastening schedule and type, foam sheathing thickness, and cladding weight.[13] When self-adhered flashing is used to treat the window openings of these assemblies, wider flashing or two overlapping pieces of flashing are required to get deep enough into the window opening (**fig. 4**). From this

FIG. 4 Two overlapping pieces of self-adhered flashing covering sill of window opening in an inverted wall.

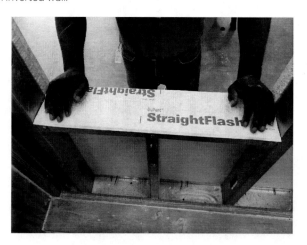

perspective, fluid applied flashing may be an easier solution at rough openings in inverted wall assemblies because the material can adapt to wider configurations more easily. However, it is important to make sure that any fluid-applied materials are compatible with the cut edges of the insulating sheathing and will not degrade the insulation over time. The flashing must be able to adhere to all substrates in the window opening, including the cut edge of the CI, regardless of whether it is a self-adhered or fluid-applied material.

Special Considerations for Specific Assemblies

CONSIDERATIONS FOR MECHANICALLY FASTENED AIR AND WATER-RESISTANT BARRIERS WITH CONTINUOUS INSULATION

Fasteners (and misplaced fasteners) are a frequent cause of water leakage in wall assemblies, especially when the systems haven not been tested together. One way to minimize this is to minimize the number of fasteners used without compromising the long-term structural performance of the assembly. When installing exterior CI over a mechanically fastened AWB, there are options to reduce the total number of fasteners penetrating the whole system. Temporary fastening of the mechanically fastened AWB can be used when the CI is installed *over* the AWB as soon as practically possible to maintain the integrity and performance of the AWB. This temporary fastening pattern adjusts for the fastening pattern used for the CI and still results in the same or more total fasteners securing the AWB for the life of the project. Regardless of the fastening pattern used, it is important to make sure that the assembly has been tested together either by the manufacturer(s) or for the project specifically. This testing will determine whether additional fastener treatment is required either over or under the CI or whether a different style of fastener should be used.

There are special considerations when foam sheathing is installed as the CI directly to the studs *under* a mechanically fastened AWB in wood-framed construction. Foam sheathing has minor nail-holding power and should not be used as the nailing base for the AWB or the exterior cladding. The mechanically fastened AWB must be fastened through the exterior CI into the framing or a layer of underlying nail-base sheathing. This may require modifying the length of the fasteners. It may also require hand nailing of the fasteners. However, similarly to when the mechanically fastened AWB is installed under the CI, it may be possible to reduce the fastener schedule of the foam sheathing itself to minimize fastener conflicts.

CONSIDERATIONS FOR THE USE OF INNOVATIVE WALL ASSEMBLIES

When changing how the control layers are sequenced from the cladding on the outside to inside of the commercial wall assembly, there are other considerations not covered in this paper that the design team must address. One example of this reordering is moving the location of the CI in relation to the AWB and cladding may change how the assembly complies with NFPA 285. Another example is altering the location of the vapor control layer will change how moisture moves moisture through the wall assembly.[14]

DESIGN AND INSTALLATION OF PUNCH WINDOW-TO-WALL INTERFACES IN WOOD-FRAMED CONSTRUCTION

Architectural design preferences and increasing continuous insulation requirements in the building code have led to changes in how punch windows must be installed to keep the exterior window in line with the exterior of the building (**fig. 5**). When the window moves out of line with the AWB, it changes how the air, water, and thermal control layers need to be detailed at the window-to-wall interface. This can sometimes contradict the preference to keep the AWB on the wall within the same plane as the air and water line of the window. This preference is based on ease of detailing and ease of installation of this transition in the field. The window-to-wall interface is especially critical to ensure that water properly drains from the window out of the assembly.

The first requirement for punch windows that must be bumped out to accommodate exterior CI and still allow the window to be in line with the exterior cladding is a bump-out frame at the rough opening (**fig. 6**).* However, integrating the

FIG. 5 Example of punch window installed such that the exterior of the window will be in line with the exterior cladding.

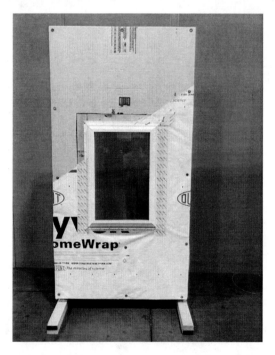

---

*Bump outs are a best practice and may not be required if the CI is 1/2 in. or less or if not required by the window manufacturer.

FIG. 6 Bump-out frame installed around rough opening.

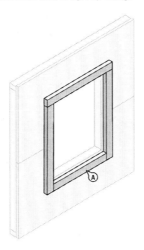

AWB from a different plane requires more complex flashing installation methods. When three-dimensional folding of self-adhered materials is required, it is important to make sure that all laps and pinholes at folds are properly sealed. Sealing of the laps and pinholes can be done using more flexible membranes, additional layers of self-adhered flashing, or with a compatible fluid-applied flashing or sealant. These details should be mocked up by the installer and tested for water penetration prior to starting work on a project. This will allow modifications to the detail to be made without requiring rework.

Fluid-applied flashing can reduce the complexity of using self-adhered flashing in complex three-dimensional details, but these products can involve material higher cost. Additionally, the installation requirements to transition between a fluid-applied flashing and a sheet-applied AWB can be challenging. Fluid-applied flashing must be confirmed to have adhesion and be compatible with the materials it is applied to and can be less forgiving on dirty substrates. Regardless of what flashing technology is chosen, it is critical that the AWB on the opaque wall be tied into the air and water line within the window system.

Detailing the window-to-wall interface must also consider the location of the continuous insulation. When installing exterior CI with an AWB, the insulation can be installed either on top of or behind the AWB. There are at least two factors to consider in the placement of the exterior insulation relative to the AWB: where the windows are located within the depth of the wall assembly and when the windows are installed in the construction sequence. As described above, locating the AWB on the same plane as the windows allows for reduced complexity of installation and an overall more robust wall system. Therefore, when possible, it is recommended that the CI move within the assembly to help the air and water control layers to remain in line while also maintaining continuous and well-placed thermal and vapor barriers. Hygrothermal analysis can be used to ensure moving the CI does not impact the water and vapor control performance of the building envelope.

To determine how best to detail different types of assemblies in wood-framed construction based on the punch window location within the plane of the wall and the order of the AWB and CI, assemblies were tested per the residential, wood-framed test protocol. Eight foot-by-eight-foot wall assemblies were built using 2" × 4" nominal wood stud and covered in oriented strand board (OSB) and included a 2' × 3' punch window opening. Multiple wall assemblies were tested to examine how best to detail the window-to-wall transition with the following variables:

1. AWB location: behind the CI, over the CI
2. Flashing type: fluid-applied flashing and self-adhered flashing, only self-adhered flashing
3. Window installation sequence: before the AWB, after the AWB

Figure 7 shows the installation sequence of the condition when the window is installed before the AWB, the CI is installed over the AWB, and self-adhered flashing is used. The window is aligned such that it would be in the same plane as the exterior cladding once it is installed. The construction and testing of the wall assemblies provided clarity on how to minimize changes from current installation guidelines when using CI. It also allowed for a demonstration that key details, such as the window corner shown in **figure 8**, will meet the building performance requirements.

When windows are installed after the AWB has been installed, it is usually possible to properly install and detail the interface regardless of the location of the CI in relation to the AWB. These installation methods are best achieved when the windows are in the same plane as the exterior cladding. When a flanged window is

FIG. 7 Window installed before the AWB and windows aligned with exterior wall: CI over AWB.

1. Install WRB.
2. Install bump out frame.
3. Protect bump out sill and window sill per current installation methods.
4. Install window.
5. Flash window.
6. Install Exterior Continuous Insulation.

FIG. 8 Corner installation detail when window is installed after AWB and the AWB is installed under CI.

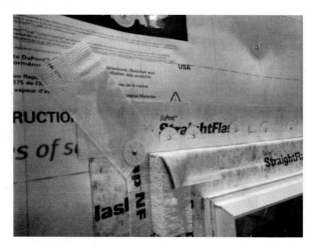

being used, a wooden bump-out frame is typically required around the window to create a solid nailing surface while at the same time keeping the outer plane of the window in line with the finished cladding. One exception to this is when the windows are recessed from the exterior cladding. When the window is recessed from the exterior cladding, it is recommended to install the AWB underneath the exterior CI. This allows for maintaining the AWB in the same plane as the window air and water line without having to work through the complex geometry of lapping the AWB back over the insulation to provide continuity with the water drainage plane of the window.

The design options for optimal installation are more limited when the windows are installed before the AWB. If the windows are installed first and recessed from the exterior wall, it is recommended to install the exterior continuous insulation on top of the AWB. This is for the same reasons cited for when the window is installed after the AWB. If the windows are installed first, and in the same plane as the exterior wall, it is recommended to install the exterior CI behind the AWB. If the windows are installed before the AWB and in the same plane as the exterior wall (**fig. 6**), a bump-out frame should be installed at the rough opening to allow for the window to be installed properly. As shown in **figure 9**, this allows for the window opening to be flashed and the window flanges to be installed directly to the AWB using the same application techniques as are detailed when there is no bump out of the window, keeping all the control layers in line. Installing the control layers in this manner reduces the complexity of installing proper flashing and integrating the AWB into the window opening. To summarize, adding bump-out framing to the rough opening allows for the proper installation of flanged and punch windows

FIG. 9 Installation sequence when window installed after AWB with window aligned with exterior cladding: AWB over CI.

1. Prepare window opening - install solid nailing surface (wood buck).
2. Install Exterior Continuous Insulation.
3. Install AWB.
4. Install flashing around window perimeter.
5. Install window.
6. Flash window flange per manufacturer recommendations.

while also providing ease of installation details in the field in the following conditions:

1. When the window needs to be aligned with exterior cladding, the AWB is installed over the CI and the window is installed before the AWB.
2. When the window needs to be aligned with exterior cladding, the AWB is installed over the CI and the window is installed after the AWB.
3. When the window needs to be aligned with exterior cladding, the AWB is installed under the CI and the window is installed before the AWB.

In other conditions, no bump-out framing is needed. While the photographs in **figure 9** show examples using mechanically fastened AWB materials, the same concepts apply to other AWB and flashing technologies.

Attaching Additional Wall Assembly Components through the AWB

ATTACHMENT OF CONTINUOUS INSULATION THROUGH THE AWB

One concern with using multiple components from multiple manufacturers when designing and constructing the whole building envelope is ensuring that the AWB remains air and water tight after the CI is installed. Additional treatment of the AWB can be required to ensure water tightness where the insulation fasteners penetrate the AWB. For fluid-applied AWBs, this can be using a thicker fluid-applied membrane or self-adhered membrane applied along the stud lines of the AWB prior to the installation of the CI. Some AWBs, particularly self-adhered membranes, do have fastener sealing properties such that they can be installed underneath the CI and still achieve high-performance air and water hold out without additional treatment. Additionally, there are fasteners that are designed to seal the AWB at the point of the fastener penetration by way of a gasket or other mechanism. There are also brick ties that double as a CI attachment that has a second gasket or seal at the AWB layer. It has been found these systems do need to be tested to confirm they meet the requirements for water infiltration for specific projects that require higher performance. Some of the fasteners can maintain air and water tightness through the assembly up to 6.24 psf, whereas others will perform up to 15 psf or higher. It has been found that whatever system or combination of systems is being used, it is important to test for water penetration at a minimum before and after structural loading to highlight any issues or confirm the use of recommended manufacturer details or fasteners. This can be completed on a project-specific basis or using pretested combinations of AWBs, CI, and fasteners.

INSTALLING GIRTS OVER CONTINUOUS INSULATION

Postinstallation penetrations are especially critical when multiple control layers are combined into a single material. Water tightness of the AWB assembly after the installation of cladding attachments, specifically rainscreen girts, is frequently questioned. To address this for foil-face ISO insulation, multiple potential solutions were tested using the established commercial wall assembly test methodology. The full study is outlined in a paper titled "Resisting Water Infiltration from Cladding Attachment Penetrations in the Wall Assembly," with key results highlighted here.[15] These tests examined different variables such as girt system type, orientations of the attachments and ways to seal the penetrations using different types of fluid and self-adhered flashing options. Numerous variables were evaluated to determine their importance as well as determine the best solution for consistency of performance:

1. ISO board thickness: 0.625"; 1.55"; 3"
2. Girt type (all continuous)—hat channel; Z-girt; proprietary horizontal system with large openings for fasteners; vertical system with large openings for fasteners
3. Girt orientation—horizontal; vertical
4. Flashing material—no flashing; silicone fluid-applied flashing; water-based acrylic flashing (A and B varieties); self-adhered flashing with a polyolefin top sheet; silyl-terminated polyether fluid-applied flashing
5. Flashing on edge—no material; top only; all*
6. Flashing on fasteners—no material; top only; all
7. Flashing in between (girt and ISO)—no material; wet material; cured material

---

*"Top only" includes only the top edge of a horizontal girt; "all" includes both top and bottom edges. This definition is only for horizontal girts because there is no physical significance of top/bottom for vertical girts. Whereas top- or bottom-edge sealing could have differences in horizontal girts, there is no expected difference between left- or right-edge sealing of a vertical girt.

The location of the flashing variations are shown in **figure 10**. The fluid-applied flashing located between the girt and the substrate was installed at the manufacturer's recommended thickness, which ranged from 15 mil wet to 45 mil wet depending on the specific flashing. Care was taken to ensure that the flashing was not squeezed out when the girts were installed when the flashing was still wet.

The testing was completed on assemblies constructed with ISO installed directly to 18-gauge steel studs spaced 16 in. on center (**fig. 11**). The seams of the ISO boards were sealed using a fluid-applied flashing solution installed per the

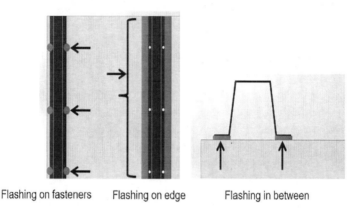

FIG. 10 Flashing locations: "on fasteners," "on edge," and "in between" shown by highlighted arrows.

Flashing on fasteners   Flashing on edge   Flashing in between

FIG. 11 Wall assembly with several girt configurations undergoing water penetration testing per ASTM E331.

manufacturer's recommendations. The study did not take into account additional variables such as the impact of the weight of cladding causing fastener rotation, the impact of thinner gauge studs, or the use of wood studs.

An early observation in the study was that insulation thickness did not have any effect on the chance of water leakage for any of the treatment scenarios. This is likely due to the flashing being installed on the surface of the ISO insulation (which serves as the AWB) and not in a separate plane from the AWB. The remaining variables studied impacted the results. The test results identified a common solution for the standard hat channel and Z-girt systems regardless of orientation: the use of the water-based acrylic flashing A cured in between the girt and the sheathing and sealing the fastener heads. The testing also provided a successful result for the proprietary attachment system, although it was different from the standard systems. This is primarily due to the large holes in that system not providing any additional gasketing or sealing at the point of fastener penetration. All passing results for each girt system are shown in **table 1**. Importantly, this research showed that these solutions are very system-specific. Every system tested passed air infiltration testing per ASTM E283 without issue. When testing for water penetration per ASTM E331 holding pressures up to 15 psf for 15 min per the commercial test protocol previously outlined, not all fluid-applied flashings performed the same. Additionally, the orientation and style of girt influenced where and under what pressures water penetration would occur. For example, a z-girt in a horizontal orientation with the silicone liquid flashing cured in between and on fasteners had water leaks starting 9 psf of pressure. However, the same assembly in the vertical orientation did not exhibit any leaks through the full protocol.

TABLE 1 Passing results—girt/flashing configuration

Girt	Flashing	Application
Hat	Silicone fluid-applied flashing	Flashing in between wet material and flashing on fasteners
	Water-based acrylic flashing A	Flashing in between cured material and flashing on fasteners
Z-horizontal	Water-based acrylic flashing A	Flashing in between cured material and flashing on fasteners
	Water-based acrylic flashing B	
Z-vertical	Water-based acrylic flashing A	Flashing in between cured material and flashing on fasteners
	Silicone fluid-applied flashing	Flashing in between wet or cured material and flashing on fasteners
Proprietary vertical attachment system	Silicone fluid-applied flashing	Wet-dipped screws
Proprietary horizontal attachment system	Silicone fluid-applied flashing	Wet-dipped screws
	Water-based acrylic flashing A	Flashing in between cured material and flashing on fasteners

The testing showed that similar solutions were successful for the standard girt systems, but greater care is required when finding a solution for proprietary girt systems. The flashing chosen had a significant impact on the results of the testing: all fluid-applied flashings do not perform the same; thicker fluid-applied membranes did not necessarily perform better than thin-film ones, and fluid-applied flashings can perform better than self-adhered membranes when sealing the fasteners of cladding attachments. Finally, the testing showed that simply sealing the edge of the girt (especially in horizontal applications) had little impact on water penetration performance through the fasteners. It was more critical to seal at the fastener penetration itself. Although the same solutions can likely be extrapolated if a WRB is installed over the ISO for a layer of redundancy, it does not apply if an AWB is installed behind the CI because the point of continuity tested is not the same.

Conclusions

The construction industry contains a diverse array of products and assemblies and an infinite combination of those products and installation methods. It can be difficult to successfully install a complete assembly with four continuous control layers, especially at the many interfaces throughout the building envelope. Research has shown that there are ways to reduce the uncertainty and achieve a high-performance building envelope. First, specify and install materials that have been tested together in the same order and with the same accessory products that will be required on the building. If the assembly needed to meet the building requirements has not been tested before, make sure that at least a mock-up is tested prior to the start of construction to work out any new details. New details could be due to the location of the AWB in relation to the continuous insulation, the location of the window in relation to the AWB, or a new accessory product designed to secure the cladding or insulation, among many other variables.

Testing assemblies before construction starts also can help reduce uncertainty in the workmanship required to properly install all the materials. The mock-up can provide a practice space for key people on the installation teams and can serve as a training area as new crew members are brought onto the job site. However, mock-ups and pretested assemblies are only as good as the interface details included. Only testing an opaque wall without penetrations does not allow for understanding the performance of the assembly at the critical interfaces. It is important to include items such as window openings, cladding attachments, and post-AWB installed penetrations, such as CI attachment. Finally, it is important to include not only air infiltration testing but also water infiltration testing and even structural and thermal stressing on the assembly if possible. This will help ensure there that a one-off detail will remain air and water tight and nothing preventable will cause damage to the building assembly over time.

ACKNOWLEDGMENTS

The testing and design of the many systems referenced in this paper were not completed by the authors alone. A huge thank you goes out to the technical teams at

DuPont and Dow that have worked on these assemblies and solutions over the years, especially to Andrew Miles, Tim Fournier, Tom Baiada, Piyush Soni, Kim LeBlanc, Shanot Kelty, and Mae Drzyzga for their more recent work on new installation methods, fastener systems, and cladding attachments.

References

1. C. Arnold, "Building Envelop Design Guide," Whole Building Design Guide, November 8, 2016, http://web.archive.org/20201218111342/https://www.wbdg.org/guides-specifications/building-envelope-design-guide/building-envelope-design-guide-introduction
2. J. Lstiburek, "Insights: The Perfect Wall," Building Science Corporation, July 15, 2010, http://web.archive.org/20210104204729/https://www.buildingscience.com/documents/insights/bsi-001-the-perfect-wall
3. J. Lstiburek, "Insight: Doubling Down—How Come Double Vapor Barriers Work?" Building Science Corporation, February 15, 2016, http://web.archive.org/20210104205030/https://www.buildingscience.com/documents/building-science-insights-newsletters/bsi-092-doubling-down%E2%80%94how-come-double-vapor-barriers
4. *Foam Plastic Sheathing Panels Used as Weather-Resistive Barriers*, ICC-ES Acceptance Criteria 71, (Brea, CA: ICC Evaluation Services, 2018).
5. C. Beall, "Heat Flow and Insulation," *Thermal and Moisture Protection Manual: For Architects, Engineers, and Contractors* (New York: McGraw-Hill, 1999), 59–120.
6. *Standard Test Method for Determining Air Leakage Rate of Air Barrier Assemblies*, ASTM E2357-18 (West Conshohocken, PA: ASTM International, approved November 15, 2018), http://doi.org/10.1520/E2357-18
7. *Standard Test Method for Determining Rate of Air Leakage through Exterior Windows, Skylights, Curtain Walls, and Doors under Specified Pressure Differences across the Specimen*, ASTM E283/E283M-19 (West Conshohocken, PA: ASTM International, approved August 1, 2019), http://doi.org/10.1520/E0283_E0283M-19
8. *Standard Test Method for Water Penetration of Exterior Windows, Skylights, Doors, and Curtain Walls by Uniform Static Air Pressure Difference*, ASTM E331-00(2016), (West Conshohocken, PA: ASTM International, approved August 1, 2016), http://doi.org/10.1520/E0331-00R16
9. *Voluntary Laboratory Test Method to Qualify Vertical Fenestration Installation Procedures*, AAMA 504-20 (Schaumburg, IL: Fenestration and Glazing Industry Alliance, 2020).
10. *Standard Practice for Determining the Effects of Temperature Cycling on Fenestration Products*, ASTM E2264-05(2013) (West Conshohocken, PA: ASTM International, approved November 1, 2013), http://doi.org/10.1520/E2264-05R13
11. *Standard Specification for Air Barrier (AB) Material or Assemblies for Low-Rise Framed Building Walls*, ASTM E1677-19 (West Conshohocken, PA: ASTM International, approved July 1, 2019), http://doi.org/10.1520/E1677-19
12. *DuPont™ Tyvek® Commercial Air Barrier Assemblies Exceed Air Barrier of America, ASHRAE 90.1 and IECC Air Leakage Requirements When Tested in Accordance with ASTM E2357* (Wilmington, DE: DuPont de Nemours, Inc., 2020), http://web.archive.org/web/20210728131728/https://www.dupont.com/content/dam/dupont/amer/us/en/performance-building-solutions/public/documents/en/tyvek-commercial-air-barrier-systems-exceed-requirements-43-d100929-enus.pdf

13. *Plastic*, 2021 International Building Code (IBC) (Country Club Hills, IL: International Code Council, Inc., 2020).26-1–26-14.
14. *Standard Fire Test Method for Evaluation of Fire Propagation Characteristics of Exterior Wall Assemblies Containing Combustible Components*, NFPA 285-19 (Quincy, MA: National Fire Protection Association, 2019).
15. A. Watts, P. Soni, and W. Su, "Resisting Water Infiltration from Cladding Attachment Penetrations in the Wall Assembly," in *Proceedings of the 2022 IIBEC International Convention and Trade Show* (Raleigh, NC: International Institute of Building Enclosure Consultants, 2022).

STP 1635, 2022 / available online at www.astm.org / doi: 10.1520/STP163520210018

Benjamin Meyer,[1] Keith Nelson,[2] and Kristin Westover[3]

Low-Slope Roofing Installations and Third-Party Observations: A Critical Review of Noncompliance Management

Citation

B. Meyer, K. Nelson, and K. Westover, "Low-Slope Roofing Installations and Third-Party Observations: A Critical Review of Noncompliance Management," in *Building Science and the Physics of Building Enclosure Performance: 2nd Volume*, ed. D. J. Lemieux and J. Keegan (West Conshohocken, PA: ASTM International, 2022), 195–217. http://doi.org/10.1520/STP163520210018[4]

ABSTRACT

Quality-assurance observations (QAOs) are an integral part of the building enclosure commissioning process. Specifically, ASTM E2947, *Standard Guide for Building Enclosure Commissioning*, includes site visits as part of the construction administration phase activities to observe the installations and compare them with the approved project documents. These observations are then reported to the project team, including photos, sketches, and noncompliances, with a method for tracking and resolving open items. For more than 10 years, more than 10,000 field observation reports have been compiled across more than 4,000 projects, resulting in more than 6,000 documented and tracked noncompliances in construction projects in more than 30 U.S. states. This large dataset will be mined to generate a collection of representative projects, focusing on new low-slope roof-system installations. Data derived from any single construction project cannot be reliably used to improve construction practices across the industry. By utilizing standard construction observation reporting procedures, such as those provided in ASTM D7186, *Standard Practice*

Manuscript received January 26, 2021; accepted for publication May 27, 2021.
[1] GAF, 5530 Fox Marsh Pl., Mosely, VA 23120, USA
[2] ECS Mid-Atlantic, LLC, 2119-D, N. Hamilton St., Richmond, VA 23230, USA
[3] GAF, 1 Campus Dr., Parsippany, NJ 07054, USA
[4] ASTM Second Symposium on *Building Science and the Physics of Building Enclosure Performance* on April 24–25, 2022, and June 12, 2022 in Seattle, WA, USA.

Copyright © 2022 by ASTM International, 100 Barr Harbor Drive, PO Box C700, West Conshohocken, PA 19428-2959.

ASTM International is not responsible, as a body, for the statements and opinions expressed in this paper. ASTM International does not endorse any products represented in this paper.

for *Quality Assurance Observation of Roof Construction and Repair*, and systematic data aggregation across a multitude of projects, this paper draws out actionable conclusions that utilize standardized QAO data collection and aggregation across projects, qualify categories of common roof-system installation noncompliance, identify trends in newly installed roof-system noncompliances and resolutions, and demonstrate the value of third-party on-site QAOs in identifying and resolving noncompliances. The research and outcomes of this paper are focused on newly installed low-slope roof-system noncompliances during construction, including their nature, prevalence, causes, and resolutions. The research and approach have not previously been published, and the results of the analysis will benefit designers, construction managers, installers, risk-management professionals, and roofing manufacturers.

Keywords
building enclosure, commissioning, BECx, quality assurance, quality-assurance observation, QAO, low-slope roof, roofing, roof inspection, site visits

Introduction

Quality-assurance observations (QAOs) are an integral part of the building enclosure commissioning (BECx) process in the construction phase and are utilized by project teams independently of the BECx process to improve roof-system installation and conformance with documented project expectations, code compliance, and manufacturer requirements. The research and outcomes of this paper are focused on construction observations of newly installed low-slope roof systems. By utilizing standardized construction observation reporting procedures and systematic data aggregation across a multitude of projects, this paper draws out actionable conclusions that utilize standardized QAO data collection and aggregation across projects, qualify categories of common roof-system installation noncompliance (NC), identify trends in newly installed roof-system NCs and resolutions, and demonstrate the value of third-party on-site QAOs in identifying and resolving NCs. The results of this analysis regarding QAO frequencies and impacts on low-slope roof nonconformance can be used to align project performance expectations, provide a framework for QAO specifications, mitigate moisture risks, and help to improve project outcomes.

Discussion

STANDARDS AND INDUSTRY PRACTICES FOR ROOF QAOS

The analysis of QAO frequencies demonstrated in this paper builds on a number of existing low-slope roof standards and industry practices. The diagram in **figure 1** roughly outlines elements of the various QAO industry resources. This analysis focuses on the QAO frequency of low-slope roofing systems, which is largely unaddressed in the current standards and industry practices.

FIG. 1 QAO standards and industry practices by specific elements.

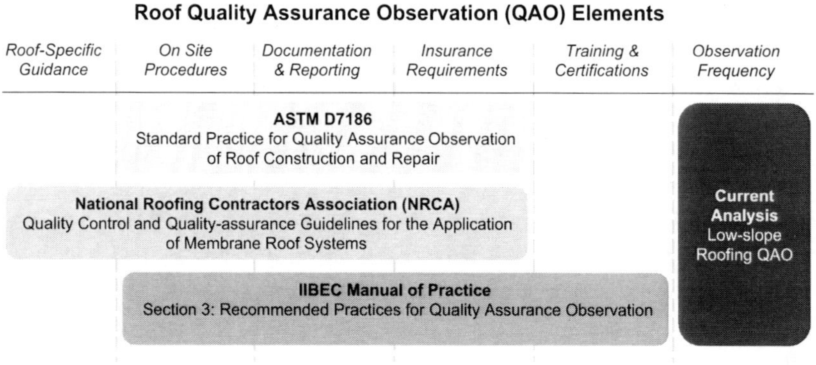

Roof Quality Assurance Observation (QAO) Elements

Roof-Specific Guidance	On Site Procedures	Documentation & Reporting	Insurance Requirements	Training & Certifications	Observation Frequency

ASTM D7186
Standard Practice for Quality Assurance Observation of Roof Construction and Repair

National Roofing Contractors Association (NRCA)
Quality Control and Quality-assurance Guidelines for the Application of Membrane Roof Systems

IIBEC Manual of Practice
Section 3: Recommended Practices for Quality Assurance Observation

Current Analysis Low-slope Roofing QAO

Roof-installation QAO practices are described in detail in ASTM D7186, *Standard Practice for Quality Assurance Observation of Roof Construction and Repair*.[1] This roofing QAO standard practice applies to new construction, roof replacements, and roof recovery but is not intended to address the condition or serviceability of existing roof systems. QAO, as defined by the standard, is "the process of recording and reporting the installation of materials and work procedures of the installer or contractor in a roofing project for the purpose of documenting compliance or non-compliance with the contract documents on a daily basis."

ASTM D7186 describes the objectives of the QAO process; the responsibilities, qualifications, and tools of the roof observer; and documentation, recording, and reporting procedures for the on-site installation or repairs. The standard goes on to describe the practice as applicable to both full-time or part-time QAO frequencies. This paper looks at the impact on low-slope roof installations with varying QAO frequencies across a large sampling of projects. It is important to note that the installation quality-control process of the installer and general contractor (GC) is separate and distinct from the QAO process.

The National Roofing Contractors Association (NRCA) quality-assurance manual[2] provides guidance for the QAO process pertaining to built-up, polymer-modified bitumen, ethylene propylene diene monomer (EPDM), and thermoplastic membrane roof systems. The NRCA document references ASTM D7186 and provides roof-type-specific guidance supplementing the observation and reporting practices in the ASTM standard. The NRCA quality-assurance manual clearly distinguishes the roles of quality control and quality assurance. Specifically, it states: "Quality control is performed by the roofing contractor" and "quality assurance, when performed, is the responsibility of the building owner's representative (e.g., architect, engineer, roof consultant) or a representative of the roof membrane manufacturer." Because the NRCA quality-assurance manual provides specific guidance based on the roof type, roofing-material manufacturers can reference the detailed

NRCA supplement to ASTM D7186. For example, a roofing-material manufacturer requires in the product-installation instructions[3] that "field quality control should be performed in accordance with NRCA's Quality Control and Quality-assurance Guidelines for the Application of Membrane Roof Systems." This requirement by the material manufacturer or the NRCA manual does not provide additional guidance regarding the frequency of QAOs for low-slope roof systems.

The International Institute of Building Enclosure Consultants (IIBEC) manual of practice[4] is a broad reference for the organization's enclosure consulting members but does include a section specifically on recommended practices for QAOs that addresses common elements across all disciplines and separate QAO discussions specific to roofs, waterproofing, and exterior walls. Although the IIBEC manual does provide a general overview of roof systems, it does not provide roof-system-specific supplemental information to perform QAOs that can be combined with ASTM D7186. IIBEC does provide a certification program to designate a registered roof observer (RRO), but neither the manual of practice nor the individual certification ensure the individual will follow the requirements in ASTM D7186 on-site. Neither the IIBEC manual of practice nor IIBEC's RRO designation provides guidance on the frequency of QAOs for low-slope roof systems.

ROOF QAOs AS PART OF THE BECx PROCESS

Two primary standards address building enclosure commissioning (BECx): ASTM E2813, *Standard Practice for Building Enclosure Commissioning*,[5] and ASTM E2947, *Standard Guide for Building Enclosure Commissioning*.[6] The diagrammatic relationship of ASTM E2813, ASTM E2947, and the current QAO analysis is shown in figure 2. This paper looks at the outcomes of a large sample of projects with

FIG. 2 Diagrammatic relationship of BECx standards and the current QAO analysis.

Building Enclosure Commissioning (BECx) Standards

ASTM E2947
Roles and tasks at each of the project phases
- Predesign Phase
- Design Phase
  - Schematic Design
  - Design Development
  - Construction Documents
- Bidding & Negotiation Phase
- Construction Phase
  - Pre-Construction
  - Construction Administration
- Occupancy & Operations Phase

ASTM E2813
Establishes two levels of BECx: Fundamental and Enhanced
- BECx Plan
- Owner Performance Requirements (OPR)
- 3rd Party Design Reviews
- Performance Testing

Current Analysis
Identifying frequencies for low-slope roofing QAO
- Fundamental
- Enhanced

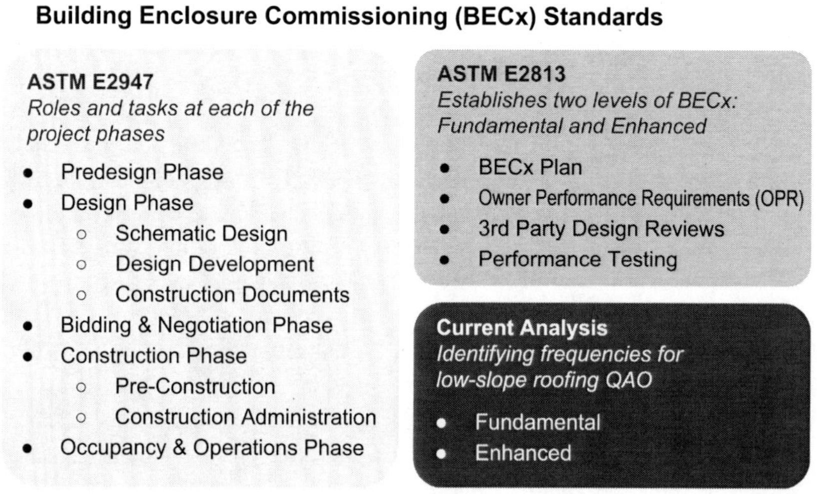

low-slope roofing installation with varying frequencies of QAOs. The analysis aims to describe the impacts and correlation of the QAO process to identify and resolve nonconforming work in roof systems.

ASTM E2813 establishes two levels of BECx: fundamental and enhanced. The standard also describes roles and responsibilities throughout the project process. For ASTM E2813, key elements of QAOs can be found in Section 5.1.4.4. The processes outlined in this section include observations, documentation, nonconformance tracking, and communication with the project team. ASTM E2947 builds on what is included in ASTM E2813 and provides additional definitions and granularity of the roles and tasks at each of the project phases.

The construction-phase QAO visits described in ASTM E2813 and E2947 verify the installed work's compliance with the construction documents and the owner's performance requirements (OPR). ASTM E2813 is flexible regarding the frequency and number of site visits and notes it "will vary based upon the complexity of the project and nature/extent of the work-in-progress and shall be determined at the sole discretion of the [Building enclosure commissioning provider] BECxP, subject to review and approval by the Owner." ASTM E2947 Section 9.3.1.2 outlines in slightly more detail the frequency and scope of installation monitoring, stating it "should be as agreed upon between the Owner and the [Building enclosure commissioning specialist] BECxS and in consultation with the [Architect of Record] AOR and BECxP" and should be performed by the BECxS be no less than at "mock-ups, commencement of new trades, benchmark installation of systems, prior to and during field testing and other scheduled times."

The lack of clarity on the number, type, and frequency of QAOs is notable for ASTM E2813, as all six pages of Annex A2 differentiate between the two levels of fundamental and enhanced BECx. Annex A2 outlines in very specific detail the minimum required tests for fundamental and enhanced BECx as defined by ASTM E2813. These detailed distinctions between the two BECx levels apply to all of the various enclosure systems but focus almost exclusively on testing methods. When a project team is selecting either fundamental and enhanced BECx levels from ASTM E2813, it is important to note that this standard does not address the appropriate frequency of QAOs for roofing systems or other enclosure elements.

ASTM E2947 provides more clarity to the BECx process and roles. For example, ASTM E2947 better defines the nonconformance process to be for "identifying, documenting, evaluating, and avoiding the inadvertent use or installation of nonconforming items of work." The QAO guidance in ASTM E2947 has slightly more information but does not differentiate between minimum project milestone QAO and a more frequent QAO installation program.

THIRD-PARTY QAO DATASET

This paper utilizes a proprietary third-party project observation database. The project information included in the database contains more than 10 years of project observations with more than 10,000 field observation reports compiled across more than 4,000 projects in more than 30 U.S. states. The database also includes the linked reports sent to the project team including photos, sketches, and NCs, along with the NC tracking and resolution log.

From this large dataset, a sampling of 55 projects that included low-slope roof-assembly QAO reports were analyzed. Not all projects in the larger dataset that included low-slope roofing were evaluated in this study. We limited the analysis sample to 55 projects because of time and resource constraints to compile and analyze the data for this paper. The 55 projects included in the analysis were initially selected to represent a range of geographical locations as shown in **figure 3**, building types as shown in **figure 4**, building heights as shown in **figure 5**, client types as shown in **table 1**, and location types as shown in **table 2**. The initial selection of projects from the larger dataset did not include screening the quantity or types of roofing NCs to avoid impacting the analysis described in the next section of this paper.

A general summary of the sampling of 55 projects with low-slope roofs for this analysis is shown in **table 3**. The 55 sampled projects represented a total of 4,346 QAO site visits; 2,961 of these were roof-related QAO visits. The projects included in the sample began after February 2014 and were completed by November 2020. These sample projects represent more than 11 million square feet (SF) of conditioned floor area and have almost 2.8 million SF of roofing. Some of the third-party

FIG. 3 Third-party QAO dataset sample population project locations.

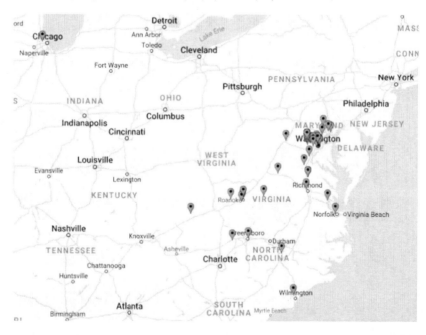

FIG. 4 Building type third-party QAO dataset sample population summary.

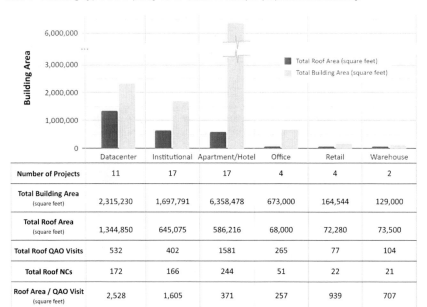

	Datacenter	Institutional	Apartment/Hotel	Office	Retail	Warehouse
Number of Projects	11	17	17	4	4	2
Total Building Area (square feet)	2,315,230	1,697,791	6,358,478	673,000	164,544	129,000
Total Roof Area (square feet)	1,344,850	645,075	586,216	68,000	72,280	73,500
Total Roof QAO Visits	532	402	1581	265	77	104
Total Roof NCs	172	166	244	51	22	21
Roof Area / QAO Visit (square feet)	2,528	1,605	371	257	939	707

FIG. 5 Histogram of building by stories in third-party QAO dataset sample project population.

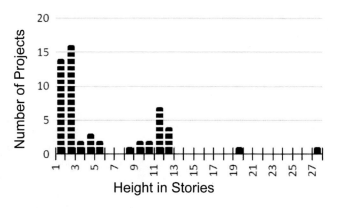

TABLE 1 Client type third-party QAO dataset sample population summary

	Owner	Developer	GC	Designer
Number of projects	23	22	8	2
Total roof NCs	205	315	136	20
Roof area/QAO visit (SF)	1,458	659	2,129	267

TABLE 2 Building location third-party QAO dataset sample population summary

	Urban	Suburban	Rural
Number of projects	22	29	4
Total roof NCs	340	319	17
Roof area/QAO visit (SF)	326	1,728	2,887

TABLE 3 Third-party QAO dataset sample population projects' summary

Total number of projects	55
Earliest project start	2/15/2014
Latest project end	11/20/2020
Total building SF	11,338,043
Total roof SF	2,789,921
Total project QAO visits	4,346
Total roof QAO visits	2,961

dataset projects included additional scopes, such as design-review services, functional performance testing of the enclosure, and enclosure commissioning tasks as described in ASTM E2813 and ASTM E2947.

The dataset and evaluation for this assessment generally exclude enclosure consulting projects that are limited in scope to forensic investigations, projects limited to the execution of specific testing procedures, and portions of projects that are limited to geotechnical, concrete, and other types of material testing. Only the low-slope roof QAO portions of the project are included in this overall assessment. Projects that were still active at the time of this assessment were also excluded from the analysis.

ROOF QAO PROJECT SAMPLE POPULATION CHARACTERISTICS

The 55 projects sampled from the third-party dataset are distributed across the Mid-Atlantic and Midwestern regions of the United States. The project locations are shown in **figure 3**. The projects of the sampled population with low-sloped roofs were sorted by building type into six primary categories: data center, institutional, apartment/hotel, office, retail, and warehouse. A high-level summary by project type is shown in **figure 4**. For this assessment, the data center, office, and warehouse project types do not have subtypes within the category. The apartment/hotel subtype includes apartments, condominiums, dormitories, and hotels. The retail building type includes fitness centers, shopping centers, and auto-repair facilities. The institutional category has the broadest subtypes and includes embassies, museums, university buildings, hospitals, fire stations, post offices, and public schools.

Within the dataset evaluated, the following roof membrane types were included in the third-party QAO sample projects:
- Thermoplastic polyolefin (TPO) single-ply
- Hot-rubberized asphalt (HRA) built-up
- Ethylene propylene diene monomer (EPDM) single-ply
- Polyvinyl chloride (PVC) single-ply
- Styrene butadiene styrene modified bituminous SBS Bit cold-applied built-up
- Polymethyl methacrylate (PMMA) liquid-applied roofing

The projects sampled from the third-party dataset with low-slope roof systems have a wide range of building heights. A histogram of the number of stories of the projects included in the sample population is shown in **figure 5**. The compiled sample population data are also sorted by client type and project location type, as shown in **tables 1** and **2**, respectively. All of the data summaries in **figure 4** and **tables 1** and **2** include the metric for the mean (average) roof area per roof QAO site visit.

The roof area per QAO site visit metric normalizes the various project total roof areas and provides an insight to the frequency of QAO visits, or, stated another way, how often a roof installation is being observed by a third party for a given roof area. For reference, if the roof area per QAO visit value decreases, this indicates that the projects are experiencing more frequent QAO visits. Looking at **table 2**, an urban project has one QAO visit for every 326 SF of roof area, whereas a rural project has one QAO visit for every 2,887 SF of roof area. This range of QAO frequency is provided for reference; additional analysis would be needed to understand the relationships between the building location, client type, and building type on NC types and frequencies.

Given this wide range of QAO frequencies, these initial summary data start to show a relationship between total roof NCs and the roof area per QAO visits. The roof area per roof QAO site visit metric is evaluated in the analysis section, in which we determine the effectiveness of QAOs in identifying NCs, resolving NCs, and testing for significant distinctions between different frequencies of low-slope roof QAOs.

QAO PROCESS AND NCs

In-progress construction QAO is an integral part of the BECx process described in ASTM E2813 and ASTM E2947. Clients can also utilize the QAO process outside of the BECx process to improve the conformance of installed enclosure systems, such as a low-slope roof assembly. The QAO data in the third-party dataset generally comply with ASTM D7186 and have been recorded and reported in a standardized observation, reporting, and compilation format across the projects discussed. **Figure 6** outlines the elements of a typical low-slope roof QAO report for the projects in the sample population dataset.

As a broad overview of the QAO site visit, the documentation and communication process are consistent regardless of QAO frequency. In a typical QAO site visit,

FIG. 6 Elements of third-party QAO reporting.

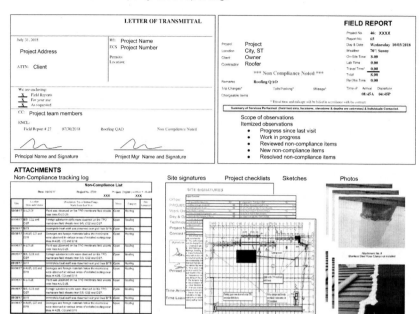

the observer is on-site for a designated time frame, generally 4–8 h, and visits the site at a frequency determined by the project team ranging from every day to on call for limited project milestones. At the end of a site visit the notable observations are discussed and signed off on-site by the responsible contractor. After the site QAO visit but before issuing the report, the documentation shown in **figure 6** is reviewed by the third-party project manager, and a second review is performed by a licensed or registered roofing professional. After the internal review is complete, the documentation is issued to the entire project team, usually within 48 h. The documentation provided to the project team with each QAO site visit generally includes the following:

- Transmittal with project information, issuing third-party information, and the enclosed report details, including if a new NC is part of the day's report.
- Field report body with a notification if a new NC is part of the day's report, time and duration of the QAO, the scope of the QAO performed, itemized observations, and current NC log items' status; NCs that have been previously resolved are not typically included in each report.
- Attachments to the report include the updated NC tracking log, site signatures, project-specific checklists, field sketches, and photographs.

In the QAO process, the goal is to identify NCs, communicate them so they can be resolved, and bring the installation into conformance with the design intent.

Ideally, a project's QAO process would identify as many available NCs as possible and then utilize the process to resolve them with the project team. Installers who are being observed with the QAO process often have a negative connotation with NCs rather than embracing the NCs as an opportunity to improve compliance with the design intent, meet project requirements, and address risks for the long-term performance of the roof. If there are too few QAO visits per roof area, there may be NCs that are present but are not captured. Likewise, if identified NCs are captured but not communicated, they may not get resolved. Moreover, if NCs are poorly tracked or not clearly communicated, this may lead to unresolved NCs. NCs that are installed but unresolved can represent future roofing risks for installers, designers, owners, insurers, or occupants. The analysis section reviews QAO frequencies and their relationship to NCs for projects.

For the NC inputs into this third-party QAO dataset, there is a consistent process for identifying, communicating, and resolving NCs for low-slope roof systems. A recorded NC is an item that is not in conformance with the project documents, manufacturer's installation instructions, or industry best practice for the system being installed. The QAO provider discusses this NC item with the installing contractor, and if the item is not resolved while the QAO provider is on-site that day, it is added to the NC tracking log. Before the QAO provider leaves the site, the responsible contractor signs a site signature sheet with the key items noted. A single NC is not repeatedly added to the NC tracking log if it is observed on multiple QAO visits, but it is noted in the field report that the item is not yet resolved. If the NC item is later corrected or accepted by the designer of record or owner, the NC tracking log is updated and reissued in a report to the project team. At the end of a project, the final NC tracking log is issued to the project team. A project team may reengage with the QAO provider to document the resolved open NC items as part of the substantial completion process, or they may perform the final resolution process directly with the responsible party. In this second case, the third-party dataset would not reflect the final status of the NC.

It is worth restating that roofing QAO does not replace the quality control required by GCs and their trade partners. Identification of a roofing NC does not only identify an isolated instance to be resolved but may also indicate a weakness in the on-site quality-control plan. When resolving an NC, it can be an opportunity to understand whether the item is isolated or systemic with the work to date.

ROOF OBSERVATION NC CATEGORIES

For this analysis of the third-party dataset, all of the NCs of the sample population projects were put in the following categories: installation, coordination, damage, storage, design, and leak. A NC in the sample population was assigned to only one category. This was possible because the specific dataset is very granular regarding the recording and nature of each NC. Categorizing the NCs allows for additional analysis of the project dataset. Below is a more detailed description of what types of NCs are present within each category.

Installation: This type of NC refers to installed low-slope roof components that are not in conformance with at least one of the following: building code, manufacturer installation instructions, industry standards, or project contract documents. These can include missing fasteners, missing accessories, improperly installed components, loose and incomplete seams, or using nonapproved alternate material.

Coordination: This type of NC is related to an installation NC because it is specific to installed material. The primary distinction for a coordination NC is the lack of coordination among trades or by the GC, or both, that led to the NC. For example, the roof system could have been initially installed in conformance, and then a subsequent contractor (e.g., plumbing, electrical, signage, scaffold, HVAC) penetrates and does not repair or coordinate the repair of the installed roof system after installing his or her components. If the NC comes as a result of designed and intended work by a subsequent contractor, it is a coordination NC type. Not included in this NC category is unintentional damage by non-roof trades on-site; these fall into the damage NC type.

Damage: This type of NC is related to the unintentional damage by roofing and non-roof trades on-site. Intentional damage by another construction trade is categorized as a coordination NC. The damage captured in this NC type applies to installed roof systems as a result of construction traffic and surrounding processes beyond the control of the installing roofing contractor. Examples of this type include unintentional damage that occurs after the roofing contractor completed the initial installation and chemical spills on the roof surface. This NC type does not include damage to materials before installation or as a result of roofing material storage conditions. This NC may include damage from material improperly stored on an installed roof that led to damage of the installed roof.

Storage: This type of NC is related to the damage and improper storage of roofing materials on-site before installation. This includes items such as not properly protecting roof-insulation bundles from the weather and storing roof materials in conditions and temperatures beyond their stated limits.

Design: This type of NC applies when the QAO provider identifies a conflict of the installed condition with the designed condition. This conflict can be related to the installer providing an unapproved alternative solution without approval, roof installation proceeding with unapproved submittals from the designer of record, and roof conditions that cannot be installed properly without design clarifications.

Leak: This type of NC applies when a roof leak is identified before the project is complete. Leaks are generally observed during construction when water is noted on the interior of the building or within a roof assembly. Leak NCs are identified on generally completed roof assemblies for the project and are noted as a stand-alone NC and are not categorized separately by their root causes, which is often not known until the leak is resolved. This NC includes visual observations of failed flood tests that are performed by the installing contractor.

Analysis

NC ASSESSMENT SUMMARY

The 55 sample projects from the third-party database had a cumulative 1,914 NCs related to these projects. Of the total number of NCs for these projects, 1,238 were not related to low-slope roofing, as shown in **figures 7** and **8**. The excluded NCs address portions of the building enclosure and structural elements that are not part of the low-slope roof assembly. The other 676 NCs were roof-related and are broken down further into the installation, damage, coordination, leak, storage, and design categories discussed above.

The top four of the six (installation, damage, coordination, and storage) low-slope roof-system NC categories combined to make up 90.1% of the NCs. The large share of these four NC types demonstrates the significance of utilizing the QAO process at a frequency that is related to the installation progress of the roof system. To identify the NC, the QAO provider needs to see the installation process before it is completed. The goal of QAO is to capture NCs as they occur, early in the installation process, and communicate them to the project team before they have the opportunity to be repeated or concealed by subsequent construction. As a result of the QAO nonconformance logging and communication process, these NCs should then be corrected and resolved to meet the project requirements.

On average, the QAO process can resolve 80.2% of the low-slope roof-related NCs identified during construction, as shown in **table 4**. When comparing the effectiveness of the QAO process, the third-party dataset can distinguish among the

FIG. 7 Third-party QAO sample population dataset NC totals.

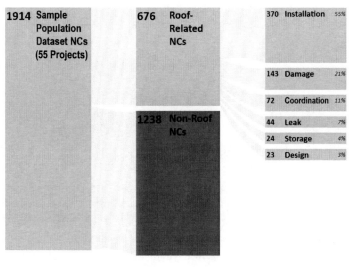

FIG. 8 Sample population dataset NC types and percentages for low-slope roof-related NCs.

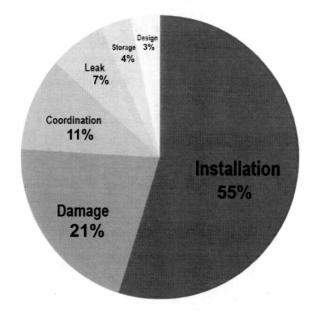

NC Type	Count	%
Installation	370	54.7%
Damage	143	21.2%
Coordination	72	10.7%
Leak	44	6.5%
Storage	24	3.6%
Design	23	3.4%
Totals	676	100.0%

TABLE 4 Third-party QAO dataset sample population NC types and resolutions for low-slope roofing

Roof NC Types	Count	Resolved		Open	
		Count	%	Count	%
Installation	370	292	78.9	78	21.1
Damage	143	118	82.5	25	17.5
Coordination	72	56	77.8	16	22.2
Leak	44	40	90.9	4	9.1
Storage	24	18	75.0	6	25.0
Design	23	18	78.3	5	21.7
All Combined	676	542	80.2	134	19.8

differences across the NC types. Material storage NCs have the lowest rate of resolution at 75%, and this may be due to the transient nature of storage conditions; by the time the third-party QAO provider returns to the site, it may not be clear or documented by the contractor what resolution was provided and whether the condition was resolved before placing the material. Roof-leak NCs identified during construction are resolved at the highest rate of all the categories at 90.9%. The high rate of resolution of a roof leak makes sense because it would be in the project team's interest to provide a performing roof system free of leaks. This does point out that the dataset still has 9.1% of the roof-leak NCs listed as open. This suggests that remediation actions are not always captured by the process.

At the end of a project, the third-party QAO provider gives the owner or project team, or both, the final QAO log. This final log is often utilized by the project team as part of the substantial completion process, and the third-party QAO provider may not be reengaged to update the ultimate status of the identified open NC items. Therefore, there are likely additional open NC items that were eventually resolved by the owner but are not reflected in the third-party database. With this logistical limitation in mind, the impact of the QAO process to resolve NCs may be undercounted in the assessment and could be considered at least as effective as shown in **table 5**.

QAO FREQUENCY GROUPING ANALYSIS

The sample population projects from the third-party database can be sorted by the roof area per roof QAO site visit metric. This metric allows for the normalization of scope and frequency across the various building factors, such as type, floor area, total roof area, location, and others. When the sample population of 55 projects is sorted, a wide range of results are present across the projects, as shown in **figure 9**. The range is as frequent as 50 square feet (SF) of roof area per roof QAO visit and as infrequent as 15,750 SF per roof QAO visit. With such a wide range of project roof QAO engagement, the sample population was broken up into three groups. Group 1 includes projects from zero to the mean, Group 2 includes projects from the mean to the mean plus one standard deviation (σ), and Group 3 includes everything greater than Group 2.

As an initial test to see whether Groups 1, 2, and 3 are different from the overall sample population, the means and 95% confidence intervals were compared visually in **figure 10**. There is a lack of overlap of the confidence intervals of Group 1 and Group 3 with the sample population. Group 2 has a mean that is different from the sample population, but there is some overlap between the 95% confidence intervals.

TABLE 5 Roof area per site visit hypothesis testing of QAO frequency groups

	Count	Mean (SF)	SD	SE	t-stat	p value	Significant?
Sample population	55	2,038	2,965	400			
Group 1	39	646	515	82	3.41	0.001	Yes
Group 2	9	2,911	785	262	−1.83	0.037	Yes
Group 3	7	8,668	3,409	1,289	−4.91	0.001	Yes

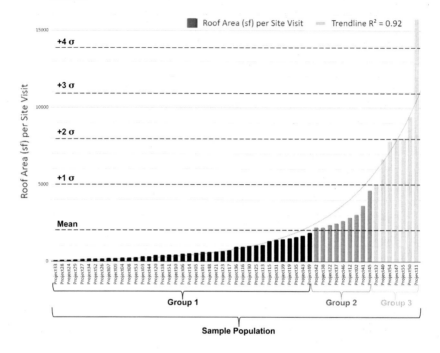

FIG. 9 Sample population sorted by roof area per roof QAO visit and grouped for analysis.

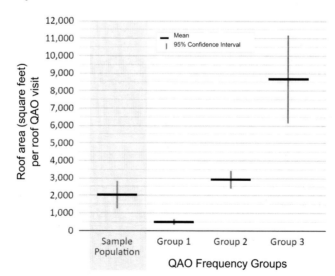

FIG. 10 Sample population sorted by roof area per roof QAO visit and grouped for analysis.

For the hypothesis testing analysis performed in **tables 5–7** the following values and terms were utilized:
- Significance occurs when the p-value (p) is less than 0.05.
- The test statistic (t-stat) compares the means and the standard error (SE) of the sample population and the group being analyzed.
- Standard deviation is represented by both σ and the abbreviation SD.

To determine whether the difference in means of the three groups is significantly different from the sample population, hypothesis testing was performed as shown in **table 5**.
- Null hypothesis (H_0): All QAO frequency groups regarding roof area per site visit are similar to the overall sample population.
- Alternate hypothesis (H_a): Not all QAO frequency groups regarding roof area per site visit are similar to the overall sample

For all three groups, the analysis shows that the groups were each significantly different than the overall population, and the null hypothesis (Ho) should be rejected and the alternate hypothesis (H_a) utilized. The results shown in **table 5** in which the p value is less than 0.05 and the alternate hypothesis should be utilized are noted as "significant," indicating a significant deviation in NC quantities.

Now that the QAO frequencies of Groups 1, 2, and 3 are determined to be distinct, the next step is to understand whether the different QAO frequency groups have an impact on the NCs compared with the overall population sample. Additional hypothesis testing was performed as shown in **table 6**.
- Null hypothesis (H_0): All QAO frequency groups (Groups 1–3) result in similar total NCs as the overall sample population.
- Alternate hypothesis (H_a): Not all QAO frequency groups (Groups 1–3) result in similar total NCs as the overall sample population.

Groups 1 and 2 were not determined to be significantly different from the overall sample population with regard to the total number of NCs for projects, so H_0 still applied. For Group 3 it shows that the quantity of NCs for this group is significantly lower than would be expected in the broader QAO sample population. Therefore, H_0 should be rejected and H_a utilized for Group 3 projects. Restated, projects in Group 3 with the lowest-frequency roof QAO site visits report significantly fewer NCs than would be expected from the sample QAO population. The results shown in **table 6** in which the p value is less than 0.05 and H_a should be utilized are noted as "significant,"

TABLE 6 Project NC quantity hypothesis testing of QAO frequency groups

	Count	Mean (SF)	SD	SE	t-stat	p value	Significant?
Sample population	55	12.3	13.2	1.8			
Group 1	39	13.5	14.4	2.3	−0.43	0.335	No
Group 2	9	12.2	11.3	3.8	0.02	0.494	No
Group 3	7	5.4	4.9	1.8	2.68	0.007	Yes

indicating a significant deviation in NC quantities. The analysis shows that there are distinct groupings of QAO frequency, and projects in the lowest-frequency Group 3 have significantly less overall NCs than the sample population.

NC CATEGORY ANALYSIS BY QAO GROUPS

The third-party dataset of NCs can be broken down further by NC categories for further evaluation. The NC categories, installation, damage, coordination, leak, storage, and design are summarized separately for the overall sample population and within each QAO group in **table 7**. By breaking the NC data up into the component categories, it is possible to see whether the QAO frequency groups have

TABLE 7 Hypothesis testing of QAO frequency groups regarding NC types and quantities

NC Category	Group	Count	Mean	SD	SE	t-stat	p value	Significant?
Installation	Sample population	55	6.7	7.1	1.0			
	Group 1	39	7.8	7.8	1.2	−0.67	0.254	No
	Group 2	5	5.1	4.4	1.5	0.92	0.185	No
	Group 3	7	3.0	3.7	1.4	2.19	0.025	Yes
Damage	Sample population	55	2.6	5.0	0.7			
	Group 1	39	2.0	3.9	0.6	0.62	0.268	No
	Group 2	9	6.0	8.9	3.0	−1.12	0.146	No
	Group 3	7	1.4	2.1	0.8	1.11	0.142	No
Coordination	Sample population	55	1.3	3.4	0.5			
	Group 1	39	1.7	3.9	0.6	−0.56	0.288	No
	Group 2	9	0.2	0.7	0.2	2.15	0.018	Yes
	Group 3	7	0.3	0.5	0.2	2.09	0.021	Yes
Leak	Sample population	55	0.8	2.5	0.5			
	Group 1	39	1.1	2.9	0.5	−0.44	0.330	No
	Group 2	9	0.2	0.4	0.1	1.59	0.059	No
	Group 3	7	0.1	0.4	0.1	1.81	0.037	Yes
Storage	Sample population	55	0.4	0.7	0.1			
	Group 1	39	0.5	0.7	0.1	−0.17	0.433	No
	Group 2	9	0.3	0.5	0.2	0.54	0.299	No
	Group 3	7	0.4	0.8	0.3	0.03	0.490	No
Design	Sample population	55	0.4	0.8	0.1			
	Group 1	39	0.5	0.9	0.1	−0.39	0.350	No
	Group 2	9	0.3	0.7	0.2	0.33	0.375	No
	Group 3	7	0.1	0.4	0.1	1.53	0.073	No

impacts on the expected number of NCs that are significantly different from the overall sample population within the NC category. Additional hypothesis testing was performed as shown in **table 7**.

- Null hypothesis (H_o): All QAO frequency groups result in similar NCs as the overall sample per NC category.
- Alternate hypothesis (H_a): Not all QAO frequency groups result in similar NCs as the overall sample per NC category.

The NC categories of installation, coordination, and leak were found to have significant differences from one or more of the QAO frequency groups than the overall sample population. At the sample population level, these three NC categories account for 71.9% of the overall NCs. Significant differences were identified in the installation NC category for Group 3 QAO frequencies, the coordination NC category for Group 2 and 3 QAO frequencies, and the leak NC category for Group 3 QAO frequencies. In these cases, H_0 should be rejected and H_a utilized. The results shown in **table 7** in which the p value is less than 0.05 and the alternate hypothesis should be utilized are noted as "significant," indicating a significant deviation in NC quantities.

In Group 3, the lowest-frequency QAO per roof area, the NC quantities for installation, coordination, and leaks are significantly below the overall sample population. This is an indication that the QAO provider is not on-site enough to witness, document, communicate, and facilitate the resolution of these categories of low-slope roofing NCs to achieve results similar to the overall QAO project sample population. These three categories of NCs, installation, coordination, and leaks, may be different from the overall population because they can be best identified by observing the installation process of a low-slope roof rather than just the final installed condition. An example of an NC that would not be observable in the final installed condition is shown in **figure 11**. Where there is no significant difference for Group 3 from the sample population, such as the damage NC category, the NCs can often be identified by inspecting final installed roof systems. The installation, coordination, and leak NC types may already be covered or concealed in the roof installation when a lower frequency in Group 3 roofing QAO is performed.

In the medium-frequency QAO group, Group 2, the ability to capture coordination NCs was found to be significantly below the overall sample population. Building construction is a dynamic process. Areas where the roofing trade has completed the roof installation are commonly penetrated or modified by subsequent installation trades. The coordination NC category is challenging because it can take an installed and conforming low-slope roof installation and make it nonconforming by proceeding further in the construction process. What these Group 2 results for coordination NCs suggest is that the highest-frequency QAO group may be needed to capture a significant amount of these dynamic NC types. The damage NC category is the largest type without significant differences between the three groups and the sample population. Damage accounts for 21.2% of the overall NC amounts. A possible reason for this consistency across categories is that damage is often visible

FIG. 11 Installation NC example. Membrane attachment out of compliance with manufacturer requirements identified while work is in progress. NC installation fasteners and plates at edge could be concealed after the subsequent roof membrane sheet installation with less frequent QAOs.

on a completed roof system. Greater-frequency Group 1 and 2 QAOs are better able to capture dynamic, installation-process-related NCs but do not lead to significantly different damage NCs of the finished low-slope roof system.

The storage and design are the two smallest NC categories, resulting in 7.0% of the sample population NCs when combined. Because they are low values for most projects, it is difficult to meet the statistical significance level of difference between the QAO groups and the sample population. What is consistent across the storage and design NC categories is that the more frequently QAO is performed, the more NCs of these types are identified. For example, **table 7** shows that the Group 1 QAO captures more storage NCs than the sample population, and Groups 2 and 3 capture equivalent or fewer NCs than the sample population. These directional results do not have enough difference, however, to categorize them as significant and reject the null hypothesis.

DISTRIBUTION OF THE DATASET POPULATION

Some extreme data points are included in the various analyses of the roof observation data. The extreme values are present in the projects' total building area, total roof area, number of observations, number of NCs, and other factors. The total range of data was included because the extreme values are not a result of errors or incorrect inclusion of non-target projects; this population data variability exists in a wide range of construction projects with low-slope roofing because real-world projects come in a wide range of shapes and sizes.

In general, the extreme data are not constrained to specific projects or specific measures; the variability is inherent in the wide range of project QAO population data. The dataset evaluated included a wide range of building types and factors that included low-slope roofing. This broad section of construction projects leads to the large variability of project QAO population data. Narrowing the analysis to specific factors, such as location types, building types, building heights, and client types would greatly reduce the variability in the data analyzed.

An example of this is the comparison of the project roof area in square feet and the number of site visits per project. As shown in **figure 12**, because of the wide range of inputs, 71% of the projects fall below the average of 2,014 SF per QAO visit. In this comparison, the standard deviation (σ) is greater than the overall average, leading to a large variance in the dataset. This wide range of results gives some of the measure distributions a largely positive skewness and a long, one-sided tail. For this reason, all of the p values presented in this paper are one-tailed p values. In addition, as a point of reference, whenever the p value was/was not significant in the results both the one- and two-tailed p values were/were not significant.

The population of buildings being evaluated is only limited by the inclusion of low-slope roofing systems, which leads to a wide range of building types and sizes. In this analysis, the range of applications for low-slope roofing included small retail buildings, multifamily buildings, large urban towers, university buildings, data centers, and warehouses. Narrowing the analysis to specific building types and client types would greatly reduce the variability in the data.

Conclusions

SUMMARY OF FINDINGS

ASTM standards for QAOs and BECx have applicable scopes for low-slope roofing and have wide ranges of implementation and results in the field. By utilizing a large

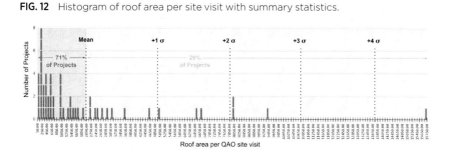

FIG. 12 Histogram of roof area per site visit with summary statistics.

standardized third-party project observation database, this analysis evaluated a sample population of projects to qualify and distinguish meaningful differences between scope and frequencies of QAO for low-slope roofing.

The differences in QAOs are primarily in the corresponding NC reporting across the NC categories of installation, coordination, damage, storage, design, and leak. Installation NCs have the highest share (54.7%), followed by damage NCs (21.2%) and then coordination NCs (10.7%). Adding in storage NCs, the top four of the six NC categories (installation, damage, coordination, and storage) are responsible for 90.1% of NCs for low-slope roof systems. As a result of the QAO process identifying, documenting, communicating, and tracking NCs, 80.2% of the low-slope roof-related NCs identified are resolved during the project.

QAO site visits per roof area is the metric used as a measure of the frequency for the project analysis. Comparing the overall sample population QAO site visits per roof area with subgroups found they are different from each other and the overall mean. Group 1 (zero to the mean), Group 2 (mean to mean $+ 1\sigma$), and Group 3 ($>$mean $+ 1\sigma$) were also compared with the overall NC rates and NC rates within each category. Regarding overall NC rates, projects in Group 3 with the lowest-frequency roof QAO site visits, report significantly fewer NCs than would be expected from the sample QAO population, indicating that NCs likely exist but remain unidentified and unresolved.

When looking at specific NC categories, significant differences were found in Group 2 and Group 3 from the overall sample population. In Group 3, the lowest-frequency QAO per roof area, the NC amounts for installation, coordination, and leaks are significantly below the overall sample population. In the medium-frequency QAO Group 2, the ability to capture coordination NCs was found to be significantly below the overall sample population.

APPLICATION OF FINDINGS

This research has identified the variability in roofing QAOs and highlights significant variation in the effectiveness of QAO frequencies in identifying and resolving NCs. The standard practice for roof-installation QAOs, ASTM D7186, does not currently distinguish between the differences and impacts of QAO frequencies. Even utilizing the BECx process as defined in ASTM E2813 and ASTM E2947 does not provide clear guidance on what level of roof QAO is appropriate for a project. The results of this research can assist in specifying the level of QAO involvement to meet the project risk profile and the OPR.

Using a similar categorization as ASTM E2813 of "fundamental" and "enhanced" to qualify roof QAO type and frequency can help distinguish between the significant results identified in this research. An enhanced level of low-slope roofing QAO would best align with the Group 1 frequency identified in the analysis in this paper, whereas a fundamental level of low-slope roofing QAO would best align with the Group 2 frequency. The Group 3 QAO frequency was found in the analysis to not align with the goals of the ASTM E2947 BECx nonconformance management process of "avoiding the inadvertent use or installation of nonconforming items of work." The Group 3 QAO frequency, greater than 5,000 SF per visit, was found to be significantly below the overall sample population for identifying NCs for installation, coordination, and leaks.

On the basis of the results of the analysis performed in this research, with the values limited to two significant figures, we propose defining low-slope roof QAO frequency in the following categories:
- Fundamental QAO: Occurs at a rate of one visit per 2,000–5,000 SF of low-slope roof area
- Enhanced QAO: Occurs at a rate one visit per 2,000 SF or less of low-slope roof area

By defining QAO frequency categories, project teams can select a QAO approach that best aligns with the size and complexity of the roof being observed. For example, a fundamental QAO frequency may be best aligned with projects that have simple roof geometries and limited roof penetrations and require minimal trade coordination. Whereas an enhanced QAO frequency may be appropriate for projects with complex roof areas, phased construction, numerous detail interfaces, and higher trade coordination expectations. The results of this analysis also show that if a project has a QAO frequency limiting observations to a roof area greater than 5,000 SF per site visit, important NCs may go unidentified and unresolved; a more frequent QAO program can be used to mitigate this risk.

Establishing a uniform specification for QAO scope and frequency would not only enhance the BECx process but also begin to remove some of the extreme variability in the overall project population. A building population with less variability could then also lead to a more refined analysis and further improvement of building enclosure design conformance.

References

1. *Standard Practice for Quality Assurance Observation of Roof Construction and Repair*, ASTM D7186-14(2021) (West Conshohocken, PA: ASTM International, approved November 1, 2021), https://doi.org/10.1520/D7186-14R21
2. National Roofing Contractors Association, *Quality Control and Quality-Assurance Guidelines for the Application of Membrane Roof Systems* (Rosemont, IL: Author, 2017).
3. GAF, *EverGuard® TPO/PVC Mechanically Attached Roofing System Overview & General Requirements Manual, Version 2.0* (Parsippany, NJ: GAF, 2019).
4. International Institute of Building Enclosure Consultants, *Manual of Practice: Roof, Waterproofing, and Exterior Wall Consulting and Quality Assurance Observation*, 3rd ed. (Raleigh, NC: Author, 2020).
5. *Standard Practice for Building Enclosure Commissioning*, ASTM E2813-18 (West Conshohocken, PA: ASTM International, approved October 1, 2018), https://doi.org/10.1520/E2813-18
6. *Standard Guide for Building Enclosure Commissioning*, ASTM E2947-16a (West Conshohocken, PA: ASTM International, approved June 1, 2016), https://doi.org/10.1520/E2947-16A

STP 1635, 2022 / available online at www.astm.org / doi: 10.1520/STP163520210032

Carly May Wagner[1] and Rex A. Cyphers[1]

Evaluation of Existing Steep Sloped Roof Assemblies for Changes to Thermal, Moisture, and Ventilation Performance: A New Joint E06/D08 Standard in Action

Citation

C. M. Wagner and R. A. Cyphers, "Evaluation of Existing Steep Sloped Roof Assemblies for Changes to Thermal, Moisture, and Ventilation Performance: A New Joint E06/D08 Standard in Action," in *Building Science and the Physics of Building Enclosure Performance: 2nd Volume*, ed. D. J. Lemieux and J. Keegan (West Conshohocken, PA: ASTM International, 2022), 218–237. http://doi.org/10.1520/STP163520210032[2]

ABSTRACT

ASTM Draft, *Standard Guide for Evaluation of Existing Steep Sloped Roof Assemblies for Changes to Thermal, Moisture, and Ventilation Performance*, is currently being developed through a joint effort between ASTM E06 and ASTM D08. The draft standard is intended to provide guidance for assessing existing steep sloped roof assemblies and attic spaces prior to altering the insulation, vapor control layers, or ventilation aspects of the assemblies. With revisions to the energy codes now requiring insulation and continuous air barriers within the building envelope, the thermal- and vapor-resistant properties of existing steep sloped roof and attic assemblies are often altered without proper evaluation of the performance of the entire enclosure. The assemblies of existing buildings must be evaluated prior to such alterations to help ensure that the addition of

---

Manuscript received March 16, 2021; accepted for publication August 4, 2021.
[1]WDP & Associates Consulting Engineers, Inc., 335 Greenbrier Dr., Charlottesville, VA 22901, USA
C. M. W. https://orcid.org/0000-0001-6913-3010, R. A. C. https://orcid.org/0000-0003-0034-7840
[2]ASTM Second Symposium on *Building Science and the Physics of Building Enclosure Performance* on April 24–25, 2022, and June 12, 2022 in Seattle, WA, USA.

Copyright © 2022 by ASTM International, 100 Barr Harbor Drive, PO Box C700, West Conshohocken, PA 19428-2959.

ASTM International is not responsible, as a body, for the statements and opinions expressed in this paper. ASTM International does not endorse any products represented in this paper.

insulation or vapor-resistant materials or the alteration to natural ventilation provisions does not cause moisture-related issues over the life of the building. This paper discusses the development of the draft standard and the strategy for formal balloting and collecting feedback between two ASTM committees. The paper also outlines the application of the standard by providing an overview of two case studies in which the procedures within the standard were executed successfully. One case study describes the field study and hygrothermal analysis needed to address condensation issues at an international airport terminal. A second case study describes the field study and review of condensation at the underside of a metal roof deck of an art museum.

Keywords

new standards, standard development, moisture content, hygrothermal analysis, vapor permeance, insulation, slate roof, metal roof, shingled roof, attic space, energy code, steep sloped roof

Introduction

In many historic buildings, the roof deck and covering served merely in a water shedding capacity and offered little to no thermal resistance or air control. Attics were naturally ventilated, and air was free to move under, around, and through the roof deck. These assemblies offer abundant drying potential for the roof deck and structural framing members but are not as energy efficient. As more and more existing and historic buildings are renovated and updated to include changes to the insulation, air barriers, vapor retarders, and mechanical systems, it is important to recognize and study the potential impacts these changes have on the existing roof and attic assemblies. Additionally, postoccupancy issues such as condensation within attic spaces, warping of wood deck materials, mold growth, decreased durability, and thermal resistance of insulation materials seem to be occurring frequently within our industry. Designers are using new insulation materials in new ways and putting mechanical systems in semiconditioned areas, often without the necessary hygrothermal analysis to confirm the design will not create a potential for condensation. Complicating the analysis is the fact that the tools commonly used for hygrothermal analysis are limited when it comes to accurately modeling naturally ventilated spaces, such as attics.

Overview of the Guide

Draft *Standard Guide for Evaluation of Existing Steep Sloped Roof Assemblies for Changes to Thermal, Moisture, and Ventilation Performance*, is intended to provide guidance for evaluating existing roof and attic spaces that are either historic or have functioned well for decades to centuries but are going to be subjected to changes that will inherently alter the conditions to which the roof and attic space are exposed or existing roof or attic spaces with existing moisture-related issues that

require alterations to address the issues. The current draft standard is aimed at evaluating the impact to existing steep sloped roof assemblies, which commonly result from:
- Addition of insulation to the attic and roof assembly;
- Alterations that impact the overall drying potential of the system, including changes to ventilation provisions or alterations to the vapor resistance of materials within the roof assembly; or
- Changes to the interior environment beneath the roof assembly, which can include:
  - Conditioning of a previously nonconditioned attic space;
  - Altering the ventilation provisions of a naturally ventilated attic space, or
  - Any other alteration that impacts the temperature, relative humidity, and air flow beneath the existing roof deck.

The current draft version of the guide is divided into the following activities:
- Review of project documents
- Determination of service history
- Visual Survey and field evaluation
- In situ data collection
- Hygrothermal analysis
- Evaluation of retrofit strategies

The current draft of the standard also features two appendices. The first is a narrative on the role of roof and attic ventilation that discusses the impact ventilation of the attic (or enclosed rafter spaces) has on the roof deck to include convective drying and temperature gradient regulation. The second appendix is a narrative on the placement of control layers and discusses the implications of a ventilated attic space versus a fully conditioned attic space versus a partially conditioned attic and the design considerations for each. It is important to note that this standard is a guide. Thus, the user of the guide should use their own engineering judgment where applicable. Further, not all activities will be feasible for all projects. The intent is to give a road map that can be adapted to suit the users' needs.

REVIEW OF DOCUMENTS AND DETERMINATION OF SERVICE HISTORY

The review of project documents and determination of service history is aimed at gathering an understanding of the existing conditions. This portion of the evaluation should seek to simply collect the information available for the existing building and answer the following questions:
- When was the building constructed?
- What is the basic design of the roof and attic assembly and has that or the mechanical design been altered since original construction?
- Are there any current known issues related to moisture? If so, when did the issues begin and under what circumstances are they present (certain seasons, times of day, weather events, all the time)?
- Are there planned alterations to the roof and attic space?

Any other basic information about the project should be amassed as needed to develop an understanding of the existing conditions.

VISUAL SURVEY AND FIELD EVALUATION
The visual survey and field evaluation includes the following activities:
- Documenting physical symptoms of moisture damage or the presence of moisture, or both, at the interior and exterior of the building to include stained, saturated, wet or damp surfaces; organic growth; corrosion of metallic elements; decay of wood deck or framing members; and areas of previous repairs or patches.
- Assessing the existing water management provisions to include a visual assessment of the elements that manage bulk water, including drainage conveyance (gutters, downspouts, drains, and scuppers) as well as typical and critical roof details and terminations (to include intersections of the roof with walls, valley flashing, expansion joints, penetrations, copings, cap flashing, and counter flashing), ridge caps, eaves and soffit vents or flashing, and any other detail that impacts how the roof manages water.
- Determination of existing construction, which consists of identifying the existing materials (type and thickness) from the interior to the exterior. This includes identifying and assessing the condition of the roof covering material, underlayment material(s), roof deck, structural supports, existing insulation, ceiling finishes, and any other material included in the roof and attic assembly.
- Determination of existing material properties consists of establishing the hygrothermal properties for each of the roofing assembly materials. Properties can be determined through testing or using published data. Properties include density, moisture storage function, vapor permeance (ideally as a function of moisture content), porosity, heat capacity, thermal conductivity, and water absorption coefficient.
- Determination of how the existing roof and attic space manages air flow includes documenting conditions that allow for natural ventilation or air exfiltration/infiltration, or both. This includes documenting the ridge and soffit vents or other ventilation design features. Additionally, smoke pencil, infrared thermography, or other methods can be used to assess passive air flow patterns.
- Diagnostic water testing is recommended if the existing roof shows signs of active or past water infiltration so that the cause of the water infiltration can be identified.

IN SITU DATA COLLECTION
This portion of the guide outlines how temperature and humidity should be recorded across the attic and roof assembly along with collecting information on the moisture content of the existing materials, surface temperatures, and differential pressures as applicable to the project. Ideally, the data collection period would occur

for a full year, but this is not always feasible given project scheduling and access. Effort should be made to collect data during both heating and cooling seasons.

HYGROTHERMAL ANALYSIS

Using the data gathered in the field, transient hygrothermal analysis should be conducted in general accordance with ASHRAE 160[1] and ASTM E3054, *Standard Guide for Characterization and Use of Hygrothermal Models for Moisture Control Design in Building Envelopes.*[2] To begin, a series of parametric analyses should be used to calibrate the model assumptions and inputs and establish the measured performance of the existing assembly. The interior temperature and humidity data collected in the field should be used for the interior climate. If sufficient data were collected, an exterior weather file can be written as well. The initial moisture contents of materials should match those measured in the field, as this helps the accuracy of the hygrothermal models. Material properties should best match those tested or gathered from published data. The heat and moisture sources and sinks should be modeled to match what was measured in the field or should be varied to demonstrate a range of conditions, noting which combination of sources and sinks result in modeled temperature and relative humidity patterns that best match those recorded in the field. It should be noted that moisture and heat transport via convection is a major shortcoming of most commonly used transient hygrothermal software packages. Thus, a range of ventilation modeling should be conducted during the process of calibrating the hygrothermal models to the measured conditions.

EVALUATION OF RETROFIT STRATEGIES

After the hygrothermal models are calibrated to best represent the existing conditions, the models should be modified to represent the planned or proposed alterations. The ultimate goal of the proposed assembly is to address any and all moisture accumulation that could result in condensation, mold growth, corrosion, structural decay of wood elements, or other failure. The standard failure criteria in ASHRAE 160 should be used with discretion. It may not be feasible to find a proposed assembly that complies with all the recommended failure criteria. However, "failing results" as deemed by ASHRAE 160 criteria should be weighed against the modeling results for the existing assembly. For example, sometimes a "do not harm" approach is appropriate. If the existing roof deck is decades old but in good condition, as long as the modeled moisture levels in the roof deck for the proposed retrofit assembly do not exceed the modeled moisture levels in the existing assembly, the results could be considered acceptable even if they exceed ASHRAE 160 criteria based on the user's judgment.

History of Development

INITIAL DRAFT AND CONCEPT

The original draft and concept of the draft standard began within E06 "Performance of Building," Subcommittee 06.24 "Building Preservation and Rehabilitation Technology."

This standard is somewhat of a companion standard to ASTM E3069, *Standard Guide for Evaluation and Rehabilitation of Mass Masonry Walls for Changes to Thermal and Moisture Properties of the Wall*,[3] in that they are rooted in the authors' experiences with retrofitting historic buildings featuring mass masonry and sloped roofs. The first drafts of these two standards date back to 2016. ASTM E3069 was quick to gather interest and was the authors' primary focus until that standard cleared balloting and was first adopted as a standard in 2017. The rough first drafts of this draft standard were polished and turned into more formal formatting throughout 2017 and 2018.

COLLABORATION WITH D08

During the first joint symposium "Building Science and the Physics of Building Enclosure Performance"[4] between D08 Roofing and Waterproofing and E06 in 2018 in Washington, DC, this draft standard gained broader attention and interest. During this time, both committees were in physical meetings at the same time and place. Until then, E06 typically met in April and October while D08 met in June and December, and in-person collaboration between the two committees had been limited.

During the committee meetings following the symposium, many D08 members attended the E06.24 subcommittee meeting, offering good feedback and interest in this draft standard. It was decided that the guide would reside under the control of E06.24 but would be balloted to both D08 and E06. The draft standard was again posted to the collaboration site, and invitations to members of D08 were sent. After limited activity on the collaboration site, the E06.24 task group decided that the draft standard was ready for balloting at the subcommittee level to gain interest and attention.

BALLOTING

In March 2020, the guide was balloted to E06.24. Six negative votes, seven affirmatives with comments, and one abstention with comment were received. All other ballots were affirmative. Comments on all ballots were constructive and are being incorporated into the current draft version of the guide.

In April 2020, the guide was balloted to D08.02 subcommittee on Steep Roofing Products and Assemblies. Five negative votes, three affirmative votes with comments, and two abstention with comments were received. All other ballots were affirmative. Comments on all ballots were constructive and are being incorporated into the current draft version of the guide.

The task group hopes to address all comments and ballot to the subcommittees in late spring 2021. The goal is to clear balloting at subcommittee by summer and ballot to committee in fall 2021. The plan is to ballot to both E06 and D08. However, comments from D08 will be considered advisory, and if the task group elects, the guide could be adopted by E06 without clearing all negative ballots from D08. That said, the task group intention is to continue to foster the collaboration between committees and to treat every D08 ballot with the same importance as each E06 ballot. The task group realizes this may mean a longer path to approval but believes that the collaboration will make for a better guide that is more widely used and recognized throughout the industry.

Case Study 1: Airport Terminal

BACKGROUND

We were contacted in 2015 by facilities managers of a mid-Atlantic international airport located within ASHRAE Climate Zone 4A. The main terminal was suffering from stained perforated metal ceiling tiles as shown in **figure 1**, which required nearly annual cleaning, which was costly due to access and disruptions to operations. Construction on the terminal was completed in 2008, and facilities managers had been working, to no avail, on adjusting mechanical systems since the completion of construction to address occupant comfort issues in the terminal.

The client provided record drawings as well as historical HVAC data, including logs of supply and return air temperature, return air relative humidity, and space temperatures. The terminal featured barrel arch construction, with a ceiling approximately 50 ft. above the finished floor at the apex. A cross-section of the terminal is shown in **figure 2**. Both the ceiling and the roof were arched, and the mechanical ducts were run within the approximately 7-ft. high, nonventilated attic space between the ceiling and the underside of the roof deck. The roof covering was a standing seam metal roof installed over 3 in. of polyisocyanurate insulation, which was installed over a vapor retarder and corrugated metal roof deck. The ceiling was a suspended perforated metal ceiling tile system.

Above each ceiling tile was acoustical insulation in the form of 1-in.-thick squares of semi-rigid mineral wool insulation within individual black polyethylene

FIG. 1 Typical staining on perforated ceiling tiles.

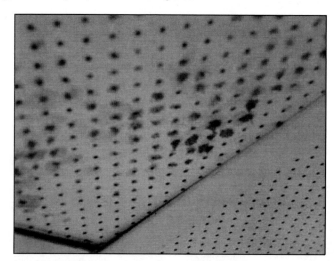

FIG. 2 Cross section of airport terminal.

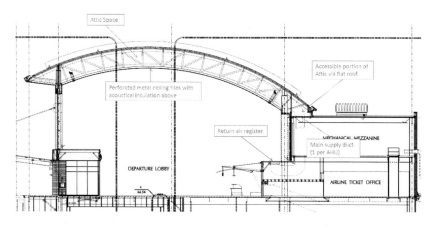

casings, as shown in **figure 3**. Each insulation pad was loose laid on each ceiling tile. Some insulation pads were missing and displaced, and air movement between and around individual insulation pads was not controlled in any manner.

DATA LOGGING

A total of 35 temperature and relative humidity data loggers were deployed throughout the attic space, exterior and interior of the terminal. The loggers were generally placed in sets stacked in a vertical fashion to gather an understanding of

FIG. 3 View of acoustical insulation installed above ceiling tiles.

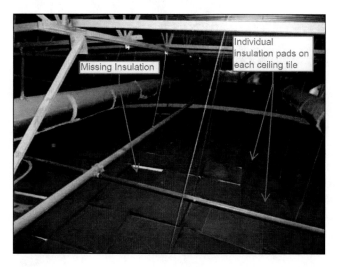

the thermal gradient of the space: one logger approximately 10 ft. above the finished floor, a second logger approximately 17 ft. above the finished floor, a third logger approximately 2 ft. below the ceiling tile, a fourth logger between the ceiling tile and the pad insulation, a fifth logger above the pad insulation, a sixth logger mid-height of the attic space, and a seventh logger within 6 in. of the underside of the roof deck. Sets of temperature and humidity loggers were placed both away from and directly near HVAC supply registers.

In addition to the temperature and humidity data loggers, thermocouples were used to record the surface temperatures of the HVAC supply registers and the metal ceiling tiles. Pressure sensors were also used to record the pressures within the occupied space as well as the attic space so that the differential pressure between the spaces could be discerned. Finally, an anemometer was used to record the wind speed at an HVAC register because, according to the mechanical engineer of record, the supply air was supposed to be provided at a high velocity so that the supply air could penetrate through the warm air within the upper parts of the arch and reach all the way down to the floor level. All logging equipment was left in place and collected data from July 2015 through March 2016.

While deploying the data loggers, a limited infrared survey was conducted to discern any obvious heat gains and losses between the attic space and the occupied space as well as between the occupied space and the exterior. The interior infrared images showed a pattern of heat gains from the attic space where the pad insulation was installed as shown in **figure 4**.

FIG. 4 Infrared image of ceiling showing heat from attic space at perimeters of the pad insulation and at missing pad insulation.

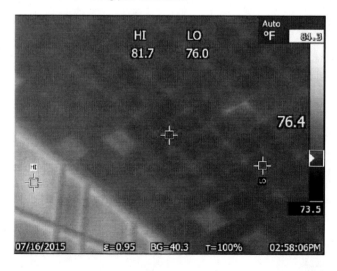

ANALYSIS

Based on an analysis of the field data collection, the following observations were made:
- The supply air velocity did not create a consistent gradation in temperature within the occupied space. Thus, the metal ceiling tiles, and especially the HVAC supply registers, did not remain consistently warmer than the air temperatures.
- Based on the pressure readings, the occupied space was under positive pressure with respect to the attic space.
- The conditioned areas of the terminal were operating under high relative humidity levels with average relative humidity of up to 65% and peak relative humidity of up to 85%.
- The average dew point temperature within the attic space was 68°F. Any duct or HVAC supply register at or below this temperature would be expected to result in surface condensation.
- The data loggers between the ceiling tiles and the pad insulation collected critical data. The dew point temperatures within this space were routinely above 70°F. Hygrothermal analysis showed that this was primarily a function of high relative humidity, as moisture between the perforated metal and the polyethylene wrapped insulation is effectively trapped here.

Transient one-dimensional hygrothermal analysis was conducted on the existing ceiling-attic-roof assembly. The software used to conduct the analysis had some drawbacks as it relates to this project. First, it is one-dimensional, so it could model the performance at the pad insulation or between the pad insulation but could not simultaneously model the impact of air movement at the edges of each pad insulation. However, when the modeled performance of the two assemblies was qualitatively considered in tandem along with the field data, it was evident how the pad insulation was trapping moisture between the perforated ceiling tiles and vapor-impermeable casing of the insulation. Subsequently, when the metal ceiling tiles dropped below ambient dew point temperature, the conditions for corrosion and biological growth on the ceiling tiles were favorable.

RECOMMENDATIONS

Between the analysis of the field study and the qualitative findings of the hygrothermal analysis, the following recommendations were provided:
- Remove all pad insulation above the ceiling tiles. This would prevent the potential for moisture accumulation between the perforated metal ceiling tiles and the impermeable casing of the insulation. This would also tend to raise the temperature of the metal ceiling tiles, decreasing the potential for the metal to drop below ambient dew point temperatures. This change alone was anticipated to prevent future systemic staining of the ceiling tiles.
- Install mechanical ventilation within the attic space to dehumidify the space and ensure all ducts and HVAC supply registers were properly insulated

and sealed with vapor-impermeable facings. This would be necessary to eliminate acute condensation at these conditions.
- Engage an HVAC consultant to work to reduce the relative humidity within the occupied space through mechanical upgrades. This would be needed to address occupant comfort complaints.

It is our understanding that the client enacted the above recommendations. We have not received any complaints or notification of ongoing or additional issues since.

Case Study 2: Art Gallery

BACKGROUND

In February 2018, we were engaged to investigate reported water infiltration at the roof of an art gallery located in West Virginia (ASHRAE Climate Zone 5A). The art gallery was originally constructed between 2013 and 2014. During the first couple of winters, reported water infiltration occurred directly below a faux dormer along the north side of the building. In 2017, the faux dormer was ultimately removed, and the north portions of the roof were replaced after unsuccessfully repairing the reported water infiltration related to the faux dormer. However, during the first winter after the reroof and dormer removal, water damage continued to occur.

Record drawings were provided for review. The art gallery included a nonventilated attic space between the gallery ceiling and the underside of the roof deck. Mechanical systems were housed within this attic space. The gallery space was relatively open to the attic space via vents and openings in the ceiling. Six inches of closed-cell spray polyurethane foam (SPF) insulation was installed at the underside of the corrugated steel structural roof deck. No insulation was installed above the steel deck. One-inch furring was installed between the plywood substrate for the asphalt shingles and the steel deck. Because the building was an art gallery, it is important to note that the interior relative humidity was kept at a constant 50%, which meant the space was humidified during the winter.

DATA LOGGING AND FIELD ASSESSMENT

It was immediately evident based on our initial visual observations that the reported water infiltration was not actually water penetration but rather condensation. **Figure 5** shows the prevalence of condensation and surface corrosion at the underside of the corrugated steel deck within an uninsulated soffit space. Nevertheless, at the client's request, diagnostic water testing was conducted to rule out any potential water infiltration that may have been concealed by condensation.

In addition to the diagnostic water testing, sets of temperature and humidity data loggers were deployed to record the conditions throughout the space at locations indicated in **figure 6**. The locations shown in section view represent one set of data loggers. A total of three sets of data loggers were deployed at various locations in plan view. Additionally, thermocouples were installed to record the temperature

FIG. 5 Condensation and corrosion on underside of existing corrugated steel deck.

FIG. 6 Section through art gallery indicating data logger locations.

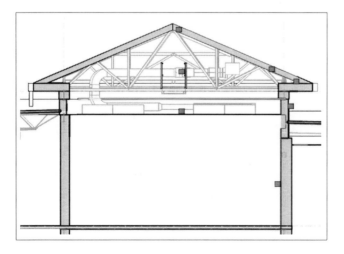

of the steel deck at the location of each set of temperature and humidity data loggers. Data were collected between early February and mid-March 2018.

Although the field data collection period was short, it was adequate to capture the condensation occurring at the underside of the steel deck. Based on the temperature and humidity recorded within the gallery space, the interior ambient dew point temperature was found to be 45°F. **Figure 7** shows the surface temperature of the steel deck as recorded by the thermocouples. At all locations, the surface temperature of the steel deck dropped below the ambient dew point temperature for several hours each day during the logging period. This meant that, if interior conditioned air was permitted to the steel deck, condensation would occur.

ANALYSIS

Transient one-dimensional hygrothermal analysis of the existing roof assembly was conducted. A series of parametric analyses were conducted to explore various conditions. In theory, closed-cell SPF insulation should function as an air and vapor barrier at the thickness installed. Thus, the first hygrothermal models did not account for any interior air migration to the underside of the metal deck. **Figure 8** shows the theoretical temperature and relative humidity at the underside of the metal deck when 6 in. of SPF insulation is installed continuously without any voids, cracks, or gaps. While this graph shows elevated relative humidity (>95%) is predicted during the

FIG. 7 Recorded surface temperatures of metal deck.

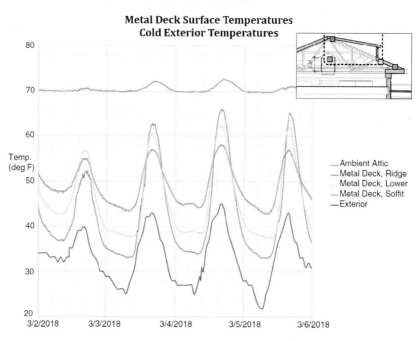

FIG. 8 Theoretical temperature and relative humidity at underside of existing steel deck.

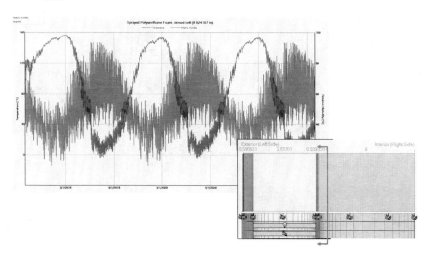

winter at the underside of the steel deck, condensation is not predicted. It should be noted that an analysis accounting for reduced thickness of SPF insulation, such as would occur at the ribs of the metal deck, was also conducted.

This theoretical modeling clearly did not match what was observed and recorded in the field. In the field, the SPF insulation was interrupted by numerous structural support members, creating thermal bridging. Additionally, there were miscellaneous voids, cracks, gaps, and discontinuities in the insulation as shown in **figure 9**. Thus, in reality, there were locations where the interior conditioned air would be permitted exfiltrate and reach the underside of the steel deck.

The initial hygrothermal models of the existing assembly were run once more with interior sourced air applied as a ventilation source to the underside of the steel deck at a rate of 1 air change per hour. When accounting to the interior air migration to the underside of the deck in this way, the models of the existing roof assembly showed condensation patterns that match those observed in the field and reported by building maintenance as illustrated in **figure 10**.

It is important to again recognize the shortcomings of the analysis software. The software is one-dimensional and does not realistically model heat and moisture via three-dimensional convection. Rather, the impacts of convection can be qualitatively approximated by using ventilation heat and moisture sources/sinks. This is also not to say that the entire roof assembly will perform as shown in either **figure 8** or **figure 10**. However, taken in tandem qualitatively and in conjunction with the field data collected, the performance of the existing roof assembly can be better understood. The root cause of the condensation was the fact that the steel deck was permitted to drop to such low temperatures. This was because, by design, virtually

FIG. 9 Cracks, thermal bridging, penetrations, and other voids in SPF as installed in the field.

FIG. 10 Temperature and relative humidity at underside of existing steel deck when accounting for air exfiltration through and around the SPF insulation.

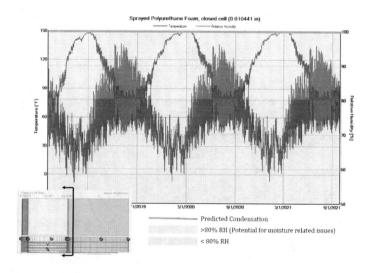

all of the thermal resistance was interior of the steel deck. Thus, the deck will remain relatively close to exterior ambient temperatures.

In addition to the transient hygrothermal analysis, steady-state finite-element two-dimensional heat-transfer simulation was conducted. The interior and exterior temperatures were selected based upon the field data collected. **Figure 11** shows the results of the thermal simulation of the existing roof assembly. It can be seen that the surface temperature of the steel deck was predicted to be 39.3°F, which is almost 6°F below the recorded interior ambient dew point temperature.

RECOMMENDATIONS

It was clear that the temperature of the steel deck would need to remain above ambient dew point because the SPF insulation installation would never be truly airtight. To accomplish this, insulation would be needed above the roof deck. The question was how much insulation would be necessary to balance the thermal gradient of the existing SPF insulation, as removal of the existing insulation would not be feasible given access and continued occupancy. Thermal simulation and transient hygrothermal models were used to assess the level of insulation needed. **Figure 12** illustrates how an additional 2 in. of polyisocyanurate insulation installed above the steel deck would raise the surface temperature of the metal deck to just above interior dew point. Similarly, **figure 13** illustrates the resulting surface

FIG. 11 Heat-transfer analysis of existing roof section showing surface temperature of corrugated steel deck below interior ambient dew point.

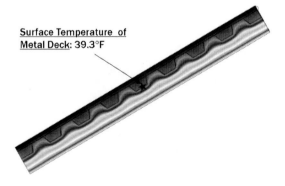

FIG. 12 Heat-transfer analysis of proposed roof section showing impact that 2 in. of polyisocyanurate insulation has on the surface temperature of corrugated steel deck, now predicted to remain slightly above ambient dew point.

Exterior: 35 °F
Interior: 65°F

6" Spray foam insulation at low flute
9" Spray foam insulation at high flute
2" Polyisocyanurate Exterior insulation

Surface Temperature of Metal Deck: 45.6°F

Color Legend
35.1° 38.8° 42.5° 46.2° 49.8° 53.5° 57.2° 60.9° 64.6°

DEW POINT OF INTERIOR CONDITIONS: 45°F

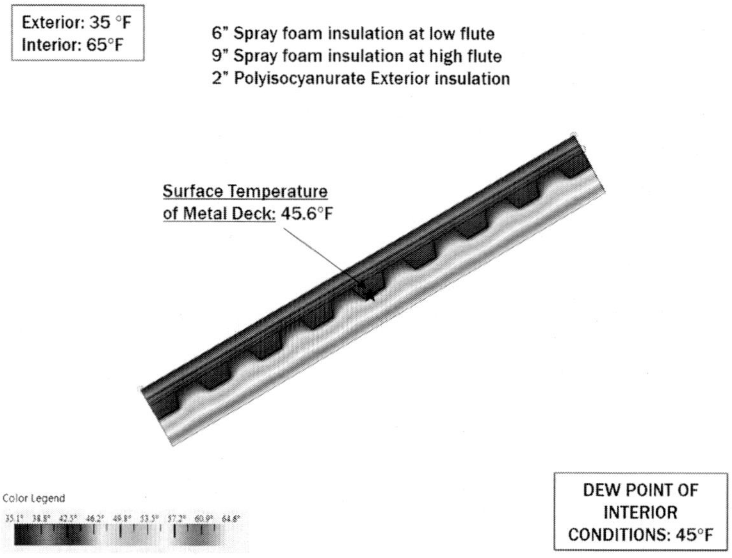

FIG. 13 Heat-transfer analysis of proposed roof section showing impact that 8 in. of polyisocyanurate insulation has on the surface temperature of corrugated steel deck, now predicted to remain above ambient dew point.

Exterior: 35 °F
Interior: 65°F

6" Spray foam insulation at low flute
9" Spray foam insulation at high flute
8" Polyisocyanurate Exterior insulation

Surface Temperature of Metal Deck: 51.8°F

Color Legend
35.1° 38.8° 42.5° 46.2° 49.8° 53.5° 57.2° 60.9° 64.6°

DEW POINT OF INTERIOR CONDITIONS: 45°F

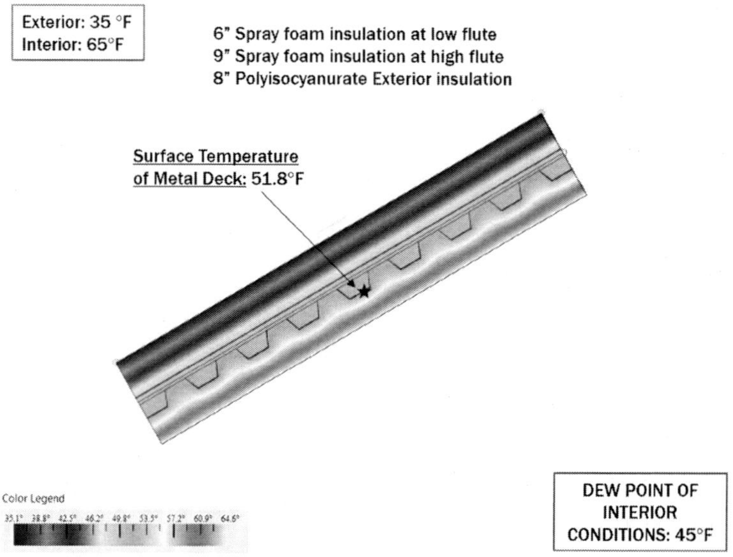

temperature assuming the recorded interior and exterior temperatures when 8 in. of polyisocyanurate insulation is installed above the steek deck. This simulation shows that the deck would remain well above the interior dew point temperature under the outlined assumptions. It should be noted that these outputs were part of a larger parametric study and focused on the temperatures and parameters recorded in the field. Ultimately, the ASHRAE Fundamentals 99.6% heating design temperature for the project site was used to inform the final recommendations.

Once again, a series of transient one-dimensional hygrothermal analyses were conducted using parameters collected in the field. The various cases explored a range of new roofing materials and insulation scenarios as well as a range of combinations of assumptions regarding heat and moisture sources and sinks. **Figure 14** illustrates the predicted temperature and relative humidity immediately below the steel deck when $5^1/_2$ in. of polyisocyanurate insulation was added above the roof deck and while accounting for the migration of interior conditioned air through and around the SPF, making contact with the steel deck. When comparing this figure to **figure 9**, which demonstrated the predicted conditions at the same location given the same moisture

FIG. 14 Predicted temperature and relative humidity at underside of steel deck when adding $5^1/_2$ in. of polyisocyanurate insulation above the steel deck and accounting for migration of interior air to the underside of the steel deck through gaps in the SPF insulation.

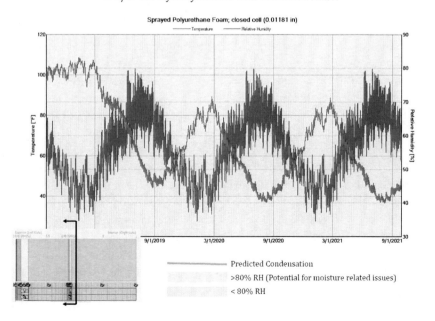

sources but under the existing assembly, it is clear that adding insulation above the steel deck would adequately mitigate the potential for moisture accumulation related issues, including condensation when it cannot be managed.

Ultimately, after optimizing the hygrothermal and thermal models, and considering various existing conditions, the following repair was enacted:
- Existing asphalt singles, plywood substrate, and furring strips were removed.
- New vapor retarder was installed on the existing steel deck.
- A new ventilated structurally insulated panel consisting of 5 in. of polyisocyanurate insulation plus a 1 in. furred air space and 3/4 in. plywood decking was installed over the vapor retarder.
- New synthetic underlayment and asphaltic shingles were installed to include a ridge vent.
- The eave/soffit details were reworked to include a continuous air barrier integrating the top of the wall to the underside of the roof deck as needed to close off the soffit space from the attic space. The new soffit included vents to serve the furred ventilation space between the insulation and the plywood roofing substrate.

The above repairs were designed and installed by October 2018. During the three winters that have passed since, there have been no reports of condensation or "water infiltration." Ultimately, the assembly described above was the third roof system a majority of the building received within approximately 4 years. This highlights the need to accurately diagnose the root cause of any moisture issue before making costly and unnecessary repairs.

Conclusions

The draft standard outlines field and analysis procedures to follow when altering how an existing roof performs. These procedures are prudent to follow when both altering an existing roof with a long history of successful performance but also for diagnosing and mitigating moisture accumulation related issues on a problematic roof or attic space. Although the task group for this draft standard falls under the control of subcommittee E06.24, coordination and input from subcommittee D08.02 has been productive and synergistic.

We have been using the methods described in the draft standard for nearly a decade on a range of projects. Two notable case studies are an international airport terminal and an art gallery. Both of these projects faced issues with moisture accumulation and condensation in locations that directly impacted the interior of the building. Field data collection and parametric hygrothermal analysis methods were used to identify the least disruptive mitigation techniques. Neither field data collection nor hygrothermal models are perfect sources in themselves because each have their own limitations. However, when considered in tandem, field data and computer models can be effectively used as a powerful design and diagnostic tool when analyzing complex roof and attic assemblies.

References

1. *Criteria for Moisture-Control Design Analysis in Buildings*, ASHRAE 160-2016 (Atlanta, GA: ASHRAE, 2016).
2. *Standard Guide for Characterization and Use of Hygrothermal Models for Moisture Control Design in Building Envelopes*, ASTM E3054/E3054M-16 (West Conshohocken, PA: ASTM International, approved March 15, 2016), https://doi.org/10.1520/E3054_E3054M-16
3. *Standard Guide for Evaluation and Rehabilitation of Mass Masonry Walls for Changes to Thermal and Moisture Properties of the Wall*, ASTM E3069-19a (West Conshohocken, PA: ASTM International, approved October 1, 2019), https://doi.org/10.1520/E3069-19A
4. D. Lemieux and J. Keegan, eds., *Building Science and the Physics of Building Enclosure Performance* (West Conshohocken, PA: ASTM International, 2020), http://doi.org/10.1520/STP1617-EB

STP 1635, 2022 / available online at www.astm.org / doi: 10.1520/STP163520200127

Travis V. Moore,[1] Mehdi Ghobadi,[1] Alex T. Hayes,[1] and Josip Cingel[1]

Experimental Thermal Resistance Comparison of a Reflective Insulation in Vertical and Horizontal Furred Airspace Orientations Using a Guarded Hot Box

Citation

T. V. Moore, M. Ghobadi, A. T. Hayes, and J. Cingel, "Experimental Thermal Resistance Comparison of a Reflective Insulation in Vertical and Horizontal Furred Airspace Orientations Using a Guarded Hot Box," in *Building Science and the Physics of Building Enclosure Performance: 2nd Volume*, ed. D. J. Lemieux and J. Keegan (West Conshohocken, PA: ASTM International, 2022), 238–251. http://doi.org/10.1520/STP163520200127[2]

ABSTRACT

Typical analyses of heat transfer across building envelopes consist of determining the thermal resistance of the assembly. The thermal resistance, or R value, is determined through test methods, calculation methods, or numerical simulations. In complex or novel wall assembly configurations, thermal resistance is required to be determined with experiments that use a guarded hot box (GHB) test apparatus according to ASTM C1363, *Standard Test Method for Thermal Performance of Building Materials and Envelope Assemblies by Means of a Hot Box Apparatus*. One scenario in which complex heat transfer occurs is in a furred airspace in contact with a low emissivity, as the combined effects of natural convection and radiation dominate the heat transfer across the space.

Manuscript received January 4, 2021; accepted for publication May 10, 2021.
[1]Construction Research Centre, National Research Council Canada, 1200 Montreal Rd., Ottawa, ON K1A 0R6, Canada
T. V. M. https://orcid.org/0000-0002-4920-9193, M. G. https://orcid.org/0000-0002-6158-9088, A.T.H. https://orcid.org/0000-0001-6333-7426
[2]ASTM Second Symposium on *Building Science and the Physics of Building Enclosure Performance* on April 24–25, 2022, and June 12, 2022 in Seattle, WA, USA.

This work is not subject to copyright law. ASTM International, 100 Barr Harbor Drive, PO Box C700, West Conshohocken, PA 19428-2959.

ASTM International is not responsible, as a body, for the statements and opinions expressed in this paper. ASTM International does not endorse any products represented in this paper.

Convection and radiation heat-transfer effects are much more sensitive to the variation in surface temperatures, orientation, and aspect ratio of the airspace. This paper presents the results of ASTM C1363 GHB tests of a full-scale (8 ft by 8 ft [2.44 m by 2.44 m]) wall assembly containing a furred airspace, with one surface having low emissivity in two configurations. The first configuration is with $3/4$-in. (19-mm)-depth strapping oriented vertically and spaced at 24 in. (0.61 m.) on center, creating four identical furred airspace cavities approximately 8 ft high by 24 in. wide. The second configuration consists of rotating the same wall assembly by 90°, creating four identical horizontally oriented furred airspace cavities. Each configuration was tested for three exterior temperature conditions: $-20°C$, $-25°C$, and $-30°C$—all with an indoor temperature of $21°C$. Additionally, the experimental results were compared with results of the ISO 6946 Annex B.2 calculation method. The results from the tests did not show a significant difference in thermal resistance results between either the exterior temperature differences or when comparing the effects of airspace orientation. This highlights some of the challenges when trying to differentiate small differences between wall assemblies with GHB testing, especially when the experimental uncertainty is considered.

Keywords

ASTM C1363, guarded hot box, radiation, natural convection, thermal resistance, *ASHRAE Handbook of Fundamentals*

Introduction

Reduction in the energy use of buildings has been identified by many countries as a critical component to meeting the greenhouse gas (GHG) emission reduction targets to limit the global temperature increase per the Paris Agreement. In North America the metric most energy codes regulate is the prescriptive minimum performance of building envelopes through specifying a minimum thermal resistance of walls and roofs for a given climate zone.[1,2] Determining the thermal resistance of a wall assembly is usually done by individually characterizing the thermal resistance of each component using a heat flow meter[3] or a guarded hot plate[4] and combining these properties using established calculation methods or numerical simulation that assumes the heat transfer through the wall assembly is dominated by conduction. However, an exception to this assumption is when the wall assembly contains a material with a low emissivity (low-e) in contact with an airspace, as the combined convective, conductive, and radiative heat transfer that occurs across the air cavity is required to be characterized. Therefore, to experimentally determine the thermal resistance of an assembly with a reflective insulation material in contact with an airspace, either field testing, guarded hot box (GHB) testing,[5] or numerical simulation accounting for the convective and radiative heat transfer across the airspace is required.

The use of low-e materials in building applications has been extensively reviewed in several publications. In general, the use of low-e materials in buildings includes radiant barriers, reflective insulation, and low-e coatings in windows.

All uses are designed to reduce the radiant heat-transfer rate across the building envelope. Jelle, Kalnaes, and Gao[6] reviewed low-e materials in use in building applications, covering both opaque (walls and roofs) and translucent (windows) applications as well as reflective insulation, radiant barriers, and low e-coatings; Tenpierek and Hasselaar[7] reviewed the use of multilayer low-e films that are separated by 2–8-mm air gaps; Yarbrough[8] and Lee, Lim, and bin Salleh[9] reviewed radiant barriers and reflective insulations; and Fricke and Yarbrough[10] reviewed methods for estimating the thermal resistance of assemblies that incorporate reflective insulation systems, including experimental, calculation, and three-dimensional numerical simulations incorporating computational fluid dynamics in the airspace. These reviews all generally show that determining the thermal resistance of a reflective insulation system (defined as a thermal insulation consisting of one or more low-e surfaces, bounding one or more enclosed air spaces) requires knowledge of the emissivity of all surfaces in contact with the airspace, the size and aspect ratio of the airspace, surface temperatures of all surfaces bounding the airspace, and the direction of the heat transfer through the airspace. Hence, it is difficult to develop general simplistic calculation methods that can work in all temperature and wall assembly cases. Therefore, typically either in situ field experiments or GHB testing is completed to develop numerical simulation calculation methods that can be used to evaluate the wall assembly in a variety of configurations.

Yarbrough[11] provided the basis for many of the following studies, including values in the *ASHRAE Handbook of Fundamentals*,[12] by correlating calculation results to several GHB tests. Kalanek, Steffek, and Ostry[13] determined the thermal resistance of representative reflective insulation samples with tests using a heat flow meter and through calculation following the method in ISO 6946. The results demonstrated that the ISO 6946 values resulted in a range of a 0.2% underestimation to a 2.8% overestimation of the thermal resistance compared with the experimental results. Pasztory et al.[14] performed an experimental investigation on the temperature-dependent thermal conductivity of a multilayer reflective insulation for mean temperatures from 0°C to 35°C, noting that the insulation system had a much stronger dependence on surface temperatures than other factors, with the thermal conductivity of the system increasing 60% from 0°C to 35°C. Additionally, increases to the emissivity of the reflective surface showed a 40% increase in thermal conductivity at 0°C when the emissivity increased from 0.058 to 0.351.

The potential for reflective insulation and radiant barrier technologies to reduce energy use in cooling environments (by reducing building heat gain due to solar radiation) has been identified as having a more significant effect than that of heating environments in which the reflective technology is used to direct heat back into the conditioned space. Fantucci and Serra[15] performed in situ experiments in Turin, Italy, comparing the ability for reflective insulation and low-e paints to reduce cooling loads in attic spaces so that they could be converted into liveable dwelling units, using the results to develop a one-dimensional (1D) calculation method to investigate other climate locations in Italy. Guo et al.[16] experimentally evaluated the effect of an exterior reflective coating on reducing the energy use of buildings under both laboratory and real climate conditions, determining that the potential for energy saving is higher for the summer months than heating months in a cooling-dominated climate location. Pourghorban, Kari, and Solgi[17] used numerical modeling in Energy Plus v7.2 and ISO 15099 to investigate the performance of reflective insulation systems in a hot-arid climate to expand the current *ASHRAE Handbook of Fundamentals* data set to higher temperature scenarios. The et al.[18] experimentally investigated the potential for using reflective insulations in below-roof applications in Southeast Asia through test huts and noted a reduction in ceiling heat flux of up to 80% in scenarios in which the reflective insulation was used.

Although increased focus has been on the use of reflective insulation and radiant barrier technologies in roofing applications in cooling-dominated climate zones, research has still occurred in heating-dominated zones. Ibrahim et al.[19] investigated the potential for a low-e coating coupled with an aerogel-based plaster on a house with no insulation in the walls through numerical modeling, comparing the effects on both the interior air temperature and operating temperature for thermal comfort. They noted that when using the interior air temperature as the control set point for the mechanical system, there was little difference in the heating load of the building. Saber[20] and Saber et al.[21] performed an analysis of reflective insulation used in a heating-dominated climate using heat flow meter and GHB tests to benchmark and validate a three-dimensional numerical simulation model that was used to develop several correlations for reflective insulations in contact with a furred, enclosed airspace. The work indicated that a heat flow meter should not be used to characterized reflective insulation systems, as it can lead to underestimations of the thermal resistance of the system due to its inability to capture the heat-transfer characteristics of the convective loops that form in the airspace. In these works, the experiment detail was provided for the heat flow meter testing such that the simulations can be repeated, including the geometry and material properties used; however, details for the GHB testing completed are not provided in sufficient detail to repeat the model.

As such, the purpose of this paper is to present the results of GHB testing of a wall assembly in sufficient detail that the results can be used to benchmark and validate numerical simulations and compare the results to the calculation method detailed in Annex B of ISO 6946. This paper presents the results of ASTM C1363, *Standard Test Method for Thermal Performance of Building Materials and Envelope Assemblies by Means of a Hot Box Apparatus*,[4] GHB tests of a full-scale (8 ft by 8 ft [2.44 m by 2.44 m]) wall assembly containing a furred airspace with one surface having low emissivity in two configurations. The wall assembly was modeled after typical basement-type installations in Canada; however, well-characterized proxy materials were used in place of the wood strapping and concrete layer to aid in future numerical modeling work. The first configuration is with $3/4$-in. (19-mm)-depth strapping oriented vertically and spaced at 24 in. (0.61 m) on center, creating four identical furred airspace cavities of approximately 8 ft in height by 24 in. in width. The second configuration consists of rotating the same wall assembly by 90°, creating a wall assembly with horizontal strapping, and furred airspaces of approximately 24 in. in height by approximately 8 ft in length. In each configuration, the wall assembly was tested for three exterior temperature conditions: −20°C, −25°C, and −30°C—all with an indoor temperature of 21°C. The test details, including geometry, material properties, and boundary conditions measured during the test, are presented in sufficient detail to enable validation of numerical simulations in follow-up work.

Materials and Methods

This section presents the details of the wall assembly, GHB test conditions, and assumptions for the ISO 6946 calculation method. The wall assembly details include the geometry specifications, material properties, and instrumentation locations. The GHB test conditions consist of the boundary conditions measured during the test and uncertainty estimate.

WALL ASSEMBLY DETAILS

GHB tests were completed at National Research Council Canada on a representative 2.44 m by 2.44 m (8 ft by 8 ft) wall assembly that consisted of, from interior to exterior, 12.7-mm-thick gypsum board; 19-mm-thick furred airspace provided by 19-mm (0.75-in.)-deep by 76.2-mm (3-in.)-wide XPS strapping spaced at 610 mm (24 in.) on center; 50 mm of foil-faced EPS (low-e facing the interior air cavity); and 15.875-mm (5/8-in.)-thick gypsum board on the exterior as a proxy for a 152-mm (6-in.)-thick concrete wall. XPS strapping was additionally placed at the top and bottom of each cavity to ensure a sealed airspace. No fasteners were used; instead, a thin layer of contact cement bonded all the layers together. As discussed previously, the wall assembly was tested in the GHB in two orientations. In the first orientation the strapping was vertical, creating air cavities that were approximately 610 mm (24 in.) wide (minus strapping) by 2.44 m (8 ft) high. The second orientation consisted of rotating the wall 90° to create horizontal furred airspaces with dimensions of 2.44 m (8 ft) wide by 610 mm (24 in.) high. Schematics of the wall assembly in both configurations are shown in **figure 1**, and photos of the as-built test assembly are shown in **figure 2**. The dimensions of each material layer are presented in **table 1**.

MATERIAL PROPERTIES

The thermal properties of the materials used in the wall assembly are presented in **table 2**. The temperature-dependent thermal conductivity of the XPS and EPS were measured using a heat flow meter in accordance with ASTM C518, *Standard Test Method For Steady-State Thermal Transmission Properties by Means of the Heat Flow Meter Apparatus*.[3] The emissivity of the reflective surface on the EPS was also measured. All other material properties were taken from the *ASHRAE Handbook of Fundamentals*.

FIG. 1 Schematic of tested wall assembly in vertical (*A*) and horizontal (*B*) orientation.

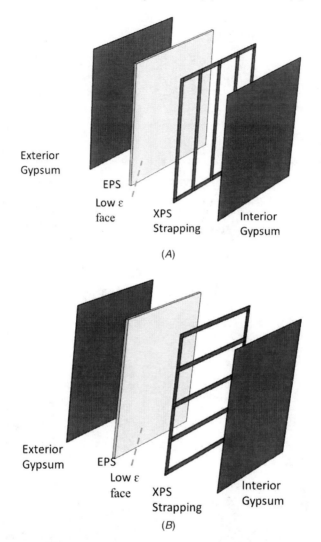

TEST CONDITIONS

The steady-state thermal resistance of the wall assembly was determined following the procedure outlined in ASTM C1363. The GHB was characterized to determine the combined metering box and flanking losses according to ASTM C1363 to ensure that the measured heat-transfer rate was that being transferred through the specimen. The combined heat-transfer coefficients presented in **table 3** are the combined effects due to radiation and convection between the specimen surfaces and the chambers on each side of the specimen and were calculated on the basis of the

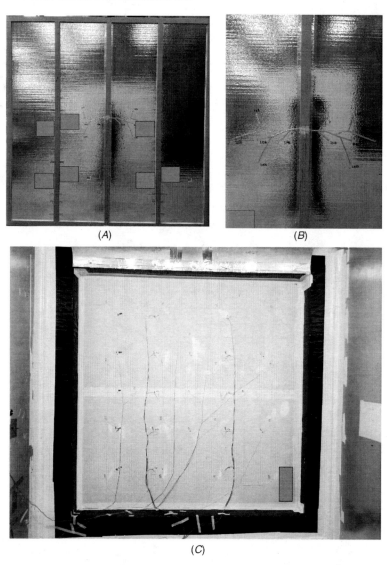

FIG. 2 Photos of foil surface (A), foil surface at temperature-sensor locations (B), and interior surface instrumentation (C).

heat flow through the wall assembly and a representative surface temperature. The ambient temperatures and film coefficients listed in **table 3** are the average air temperature measured during the tests on the interior and exterior. The uncertainty of the results is estimated at ±9% for the thermal resistance and ±0.7°C (33.2°F) for the temperatures; the uncertainty includes both the uncertainty of the measurement

TABLE 1 Wall assembly dimensions

Material Layer	Length mm (in.)	Width mm (in.)	Thickness mm (in.)
Exterior gypsum sheathing (concrete proxy)	2,438.4 (96)	2,438.4 (96)	15.875 (0.625)
XPS Strapping[a,b,c]	2,438.4 (96)	76.2 (3)	19 (0.75)
EPS	2,438.4 (96)	2,438.4 (96)	50.8 (2)
Interior gypsum sheathing[a]	2,438.4 (96)	2,438.4 (96)	12.7 (0.5)

[a]dimensions provided for vertical cavity orientation, rotate 90° for horizontal cavity orientation
[b]horizontally spaced at 609.6 mm (24 in.) on center
[c]strapping also horizontally at the top and bottom of each cavity to create enclosed space

TABLE 2 Material properties

Material	Density, kg/m^3 (lb/ft^3)	Specific Heat, J/(kg × K) (BTU/lb × °F)	Thermal Conductivity, W/mK (BTU/h × ft°F)	Emissivity
Gypsum	700 (43.7)	870 (0.208)	0.159 (0.0919)	0.9
XPS strapping	22 (1.37)	1,500 (0.358)	$k(T) = 0.026 + 0.0001129 \times (T[°C], (5/9*(T[°F] - 32))$	0.9
EPS	14.8 (0.92)	1,470 (0.351)	$k(T) = 0.000129(T[°C], (5/9 \times (T[°F] - 32))) + 0.034$	0.0351

TABLE 3 GHB test conditions

Boundary Condition	Film Coefficient, W/m^{2}K (BTU/h*ft^2°F)	Ambient Temperature, °C (°F)
Interior surface	5.5 (0.969)	20.67 (69.21)
Exterior surface	13 (2.29)	−19.48 (−3.06), −24.39 (−11.88), −29.30 (−20.74)
Lateral sides	0	

devices as well as variations in the readings of the sensors during the measurement. It is not stated in ISO 6946 what the uncertainty of the calculation method might be—however, for the purposes of this paper it is estimated to be the same as the experimental uncertainty.

ISO 6946 CALCULATION ASSUMPTIONS

The calculation method used to compare with the GHB tests for the wall assembly described in this paper was ISO 6946 Annex B.2, as the air gap consists of a closed airspace in contact with a low-e surface where the length and width of the airspace are both greater than 10 times the depth of the cavity. To calculate the equivalent thermal resistance of the cavity according to ISO 6946, two input values are required: the temperature difference across the airspace and the mean temperature of the airspace. In most situations outside of a lab test, exact knowledge of these two values needs to be calculated. In this work, these values were estimated by assuming a 1D heat-transfer path shown in the resistance diagram in **figure 3** and solving for the surface temperatures on each side of the airspace (T_2 and T_3) using assumed values and an iterative calculation procedure.

The iterative calculation procedure consists of determining the heat flux per unit area through the wall by assuming that the material properties of all solid materials, interior and exterior ambient temperatures, and interior and exterior film coefficients are known. The initial thermal resistance of the airspace is assumed to start the iterative process (a nominal thermal resistance of $0.76\,\text{m}^2\text{K/W}$ was assumed corresponding to a thermal conductivity of $0.025\,\text{W/mK}$). The film coefficients are taken from the *ASHRAE Handbook of Fundamentals*, with the interior as $8.33\,\text{W/m}^2\text{K}$ and the exterior as $33.33\,\text{W/m}^2\text{K}$. These were used in lieu of the GHB test result because they are more likely to be assumed in a calculation scenario outside of a lab test. The interior and exterior ambient temperatures were taken from the GHB test. The heat flux is solved for by determining the total thermal resistance as shown in equation (1), combined with the temperature difference between the ambient airspaces using equation (2).

$$R_{total} = \sum_{in}^{out} R = \frac{1}{h_{in}} + \frac{l_1}{k_1} + \frac{1}{h_a + h_r} + \frac{l_2}{k_2} + \frac{l_3}{k_3} + \frac{1}{h_{out}} \quad (1)$$

where:

h_{in} = interior film coefficient,
l_1 = gypsum thickness,
k_1 = thermal conductivity of gypsum,
h_a = ISO 6946 convective coefficient of the reflective airspace,
h_r = ISO 6946 radiative coefficient of the reflective airspace,
l_2 = EPS thickness,
k_2 = thermal conductivity of EPS,

FIG. 3 Heat-transfer resistance diagram for ISO 6946 calculations.

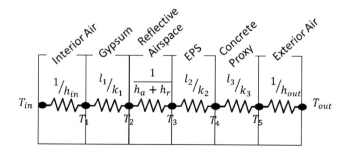

l_3 = proxy concrete thickness,
k_3 = thermal conductivity of gypsum, and
h_{out} = exterior film coefficient.

$$q = \left(\frac{1}{R_{total}}\right) * (T_{in} - T_{out}) \qquad (2)$$

where:
q = heat flux, W/m²,
R_{total} = total thermal resistance calculated from **figure 3**,
T_{in} = interior air temperature, and
T_{out} = exterior air temperature.

Thereafter, the heat flux is used with the thermal resistance equation to determine the surface temperatures at T_2 and T_3, which correspond to the surfaces in contact with the airspace; T_2 is the interior surface of the gypsum board, and T_3 is the low-e surface. The surface temperatures at T_2 and T_3 are calculated using equation (3) and equation (4), respectively.

$$T_2 = \left(T_{in} - q * \sum_{in}^{T_2} R\right) = \left(T_{in} - q * \left(\frac{1}{h_{in}} + \frac{l_1}{k_1}\right)\right) \qquad (3)$$

$$T_3 = \left(T_{out} + q * \sum_{T_3}^{out} R\right) = \left(T_{in} + q * \left(\frac{l_2}{k_2} + \frac{l_3}{k_3} + \frac{1}{h_{out}}\right)\right) \qquad (4)$$

Once the surface temperatures were determined they were used to determine the mean temperature and temperature difference across the airspace. The mean temperature and temperature difference were then used with the correlations from ISO 6946 Annex B.2 to determine the effective thermal resistance of the airspace using equations (5)–(9).

$$R_{airspace} = \frac{1}{h_a + h_r} \qquad (5)$$

$$h_a = 0.73 * (T_2 - T_3)^{\frac{1}{3}} \qquad (6)$$

$$h_r = E * h_{r0} \qquad (7)$$

$$E = \frac{1}{\frac{1}{\epsilon_1} + \frac{1}{\epsilon_2} - 1} \qquad (8)$$

$$h_{r0} = 4 * \sigma * \left(\frac{T_2 - T_3}{2}\right)^3 \qquad (9)$$

The effective thermal resistance for the airspace was then plugged back into equation (1) to solve for the new total thermal resistance, which was used in equation (2) to solve for the new heat flux, which was then used to resolve the surface temperatures at T_2 and T_3 (equation (3) and equation (4), respectively), which were used to solve for the new effective thermal resistance of the airspace using equations (5)–(9). This cycle was repeated until the heat flux converged to within two decimal places. This calculation method does not differentiate between the horizontal and vertical cavity orientation, so only one set of calculation values is presented.

Results and Discussion

The results of the GHB tests and corresponding ISO 6946 calculation results for each orientation and temperature are presented in **figure 4** and **table 4**, respectively. As can be seen from the results presented in **figure 4**, there is very little difference between the results for each orientation for each different exterior temperature, and, in fact, the variations between orientation and test are largely within the uncertainty estimated for the GHB test results.

FIG. 4 Thermal resistance results for the vertical and horizontal cavity orientations at each exterior temperature.

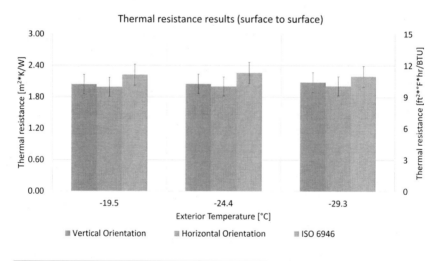

TABLE 4 Thermal resistance results for the vertical and horizontal cavity orientations at each exterior temperature.

Temperature (°C)	Vertical Orientation, RSI (m²K/W)	Horizontal Orientation, RSI (m²K/W)	ISO 6946, RSI (m²K/W)
−19.5	2.04	1.99	2.22
−24.4	2.05	2.00	2.26
−29.3	2.08	2.01	2.20

Comparing the results of the vertical orientation and the horizontal orientation to the results from the ISO 6946 calculation method demonstrates that the ISO 6946 method results in an slight overestimation of the thermal resistance; however, the results are very close to the uncertainty bands of the experiment results. For the vertical orientation, the results have a maximum deviation of 10%, whereas for the horizontal orientation it is slightly higher at a maximum deviation of 13%. Given that the calculation assumptions used in this paper for the ISO 6946 calculation used material properties at the overall average air-to-air temperature and a 1D heat-transfer case was assumed, the results of the calculation were considered within good agreement of the experiment results.

Overall, the results highlight some of the difficulties in evaluating the performance of systems with reflective insulation systems. Although differences were found between the vertical and horizontal orientations, and slight differences with temperature, it is difficult to comment on how the reflective insulation system versus the temperature dependence of the materials contributed to the differences. Additionally, all differences in the results were within the uncertainty of the GHB results, adding another difficulty when trying to make comparisons. A further comparison of the contribution of these systems to the overall R value will have to be made using the numerical simulation software.

Conclusions and Future Work

This paper experimentally determined the thermal resistance of a wall assembly that contained a reflective insulation in contact with a furred enclosed airspace using a GHB test. The wall assembly was evaluated for two airspace orientations, a vertical orientation and a horizontal orientation, achieved by rotating the wall assembly 90° between tests. Each orientation was evaluated for three different exterior temperature conditions: $-19.5°C$, $-24.4°C$, and $-29.3°C$. The paper additionally presented the material properties, geometry, and GHB test conditions in sufficient detail to enable validation of numerical simulation software in future work. The experimental results were also compared with results from the ISO 6946 Annex B.2 calculation method.

Comparing the thermal resistance results between the orientations and across the three different exterior temperatures did not yield much difference in the results, especially when the uncertainty of the GHB test is considered. This highlights some of the issues that can occur when testing potentially small differences between wall assemblies in a GHB. Additionally, the difference between the ISO 6946 calculation method and the experimental results showed a slight overestimation in the thermal resistance; however, given the uncertainty of the experimental and calculation method, these results could be considered within good agreement of the experiment results.

The work provided in this paper should be sufficient to validate numerical simulations, which can be used to differentiate the effects of different components' contribution to the overall thermal resistance of the wall assembly. Additionally, the numerical work can be extrapolated to other temperature conditions.

ACKNOWLEDGMENTS

We thank Michael Nicholls and Michael Ryan of National Research Council Canada for their efforts in the construction of the wall assembly. This work was funded by National Research Council Canada.

References

1. *Energy Standard for Buildings Except Low-Rise Residential Buildings*, ANSI/ASHRAE/IES Standard 90.1-2016 (Atlanta, GA: ASHRAE, 2016).
2. National Research Council Canada, *National Energy Code of Canada for Buildings* (Ottawa: Author, 2016).
3. *Standard Test Method For Steady-State Thermal Transmission Properties by Means of the Heat Flow Meter Apparatus*, ASTM C518-17 (West Conshohocken, PA: ASTM International, approved May 1, 2017), https://doi.org/10.1520/C0518-17
4. *Standard Test Method for Thermal Performance of Building Materials and Envelope Assemblies by Means of a Hot Box Apparatus*, ASTM C1363-11 (West Conshohocken, PA: ASTM International, approved May 15, 2011), https://doi.org/10.1520/C1363-11
5. *Standard Test Method for Steady-State Heat Flux Measurements and Thermal Transmission Properties by Means of the Guarded-Hot-Plate Apparatus*, ASTM C177-19 (West Conshohocken, PA: ASTM International, approved January 1, 2019), https://doi.org/10.1520/C0177-19
6. B. P. Jelle, S. E. Kalnaes, and T. Gao, "Low-Emissivity Materials for Building Applications: A State-of-the-Art Review and Future Research Perspectives," *Energy and Buildings* 96 (2015): 329–356.
7. M. Tenpierik and E. Hasselaar, "Reflective Multi-Foil Insulations for Buildings: A Review," *Energy and Buildings* 56 (2013): 233–243.
8. D. Yarbrough, "Reflective Materials and Radiant Barriers for Insulation in Buildings," in *Materials for Energy Efficiency and Thermal Comfort in Buildings*, ed. M. R. Hall (Woodhead Publishing Limited: Cambridge, UK, 2010), 305–318.
9. S. W. Lee, C. H. Lim, and E. bin Salleh, "Reflective Thermal Insulation Systems in Building: A Review on Radient Barrier and Reflective Insulation," *Renewable and Sustainable Energy Reviews* 65 (2016): 643–661.
10. J. Fricke and D. Yarbrough, "Review of Reflective Insulation Estimation Methods" (paper presentation, 12th Conference of International Building Performance Simulation Association, Sydney, Australia, November 14–16, 2011).
11. D. Yarbrough, "Assessment of Reflective Insulations for Residential and Commercial Applications," Technical Report TM-8891 (Oak Ridge, TN: Oak Ridge National Laboratory, 1983).
12. ASHRAE, *ASHRAE Handbook of Fundamentals* (Atlanta, GA: Author, 2016).
13. J. Kalanek, L. Steffek, and M. Ostry, "Compare of Experimental and Numerical Evaluation of Structure Application of Reflective Insulation," *Applied Mechanics and Materials* 824 (2016): 34–41.
14. Z. Pasztory, T. Horvath, S. Glass, and S. Zelinka, "Experimental Investigation of the Influence of Temperature on Thermal Conductivity of Multilayer Reflective Thermal Insulation," *Energy and Buildings* 174 (2018): 26–30.

15. S. Fantucci and V. Serra, "Investigating the Performance of Reflective Insulation and Low Emissivity Paints for the Energy Retrofit of Roof Attics," *Energy and Buildings* 182 (2019): 300–310.
16. W. Guo, X. Qiao, Y. Huang, M. Fang, and X. Han, "Study on Energy Saving Effect of Heat-Reflective Insulation Coating on Envelopes in the Hot Summer and Cold Winter Zone," *Energy and Buildings* 50 (2012): 196–203.
17. A. Pourghorban, B. M. Kari, and E. Solgi, "Assessment of Reflective Insulation Systems in Wall Application in Hot-Arid Climates," *Sustainable Cities and Society* 52 (2020): 101734.
18. K. S. The, D. Yarbrough, C. H. Lim, and S. Salleh, "Reflective Insulation for Energy Conservation in South East Asia," *Earth and Environmental Science* 62 (2017): 1–6.
19. M. Ibrahim, L. Bianco, O. Ibrahim, and E. Wurtz, "Low-Emissivity Coating Coupled with Aerogel-Based Plaster for Walls' Internal Surface Application in Buildings: Energy Saving Potential Based on Thermal Comfort Assessment," *Journal of Building Engineering* 18 (2018): 454–466.
20. H. Saber, "Investigation of Thermal Performance of Reflective Insulations for Different Applications," *Building and Environment* 52 (2012): 32–44.
21. H. H. Saber, W. Maref, M. C. Swinton, and C. St-Onge, "Thermal Analysis of Above-Grade Wall Assembly with Low Emissivity Materials and Furred-Airspace," *Building and Environment* 46 (2011): 1403–1414.

Kyle H. Jang,[1] Elyse A. Henderson,[1] and Graham Finch[2]

Moisture Management in Wood Volumetric Modular Construction

Citation

K. H. Jang, E. A. Henderson, and G. Finch, "Moisture Management in Wood Volumetric Modular Construction," in *Building Science and the Physics of Building Enclosure Performance: 2nd Volume*, ed. D. J. Lemieux and J. Keegan (West Conshohocken, PA: ASTM International, 2022), 252–268. http://doi.org/10.1520/STP163520210031[3]

ABSTRACT

Prefabricated volumetric modular construction is an approach used by housing providers to quickly provide affordable, temporary, and permanent housing to remote or underserved communities. Modular construction presents unique circumstances for moisture management that conventional construction does not typically encounter, and this is especially important in wet climates. Modular buildings have inherent building characteristics, such as interstitial spaces, that can lead to a higher risk of moisture accumulation. This paper highlights typical sources of moisture in modular construction (providing details of water ingress into modular units during transportation, storage, complexing, and postoccupancy), outlines moisture protection measures throughout all stages of the project, and provides a summary of recommendations for moisture management. It is recommended that moisture management planning, including a robust risk analysis, and the development of a moisture management plan become a standardized process throughout the modular construction industry. The development of a standard or standards could improve the process of delivering dry modules to project sites and improve the quality, consistency, and confidence of modular construction.

Manuscript received March 16, 2021; accepted for publication July 26, 2021.
[1] RDH Building Science Inc., 400-4333 Still Creek Dr., Burnaby, BC V5C 6S6, Canada
K. H. J. http://orcid.org/0000-0002-7116-995X, E. A. H. http://orcid.org/0000-0002-2079-4974
[2] RDH Building Science Inc., 602-740 Hillside Ave., Victoria, BC V8T 1Z4, Canada
[3] ASTM Second Symposium on *Building Science and the Physics of Building Enclosure Performance* on April 24-25, 2022, and June 12, 2022 in Seattle, WA, USA.

Copyright © 2022 by ASTM International, 100 Barr Harbor Drive, PO Box C700, West Conshohocken, PA 19428-2959.

ASTM International is not responsible, as a body, for the statements and opinions expressed in this paper. ASTM International does not endorse any products represented in this paper.

Keywords

modular construction, moisture management, membranes, critical barriers, modular, prefabricated, volumetric

Introduction

Prefabricated volumetric modular construction is an approach that is increasingly being used by housing providers to quickly provide affordable, temporary, and permanent housing to remote or underserved communities.[1,2] Modular construction is also used widely for multifamily residences, hotels, and office buildings. Modular construction can be useful where housing is required to be built with a shorter on-site construction time period, where a large amount of housing is required in a short time period, in areas in which there is a lack of skilled workers available or lack of building supplies, and for sites that are only available on a temporary basis. The term "prefabrication" includes both panelized and volumetric construction. Volumetric modular construction refers to three-dimensional closed or open modules.[3] This paper focuses on closed volumetric modules (one module per unit with connections to other modules at doors, shafts, etc.) versus open modules (multiple modules per unit with connections at open walls or roofs).[3] The modules are constructed in a manufacturing facility and then transported and assembled on site. **Figure 1** shows a typical module being craned into place during the on-site assembly of a modular building.

Moisture management is an important consideration in any construction project. Poor indoor air quality due to fungal growth and diminished service life due to

FIG. 1 Typical closed module with six exposed sides of the volume during transport and installation and terminology for the typical geometry and protection strategies.

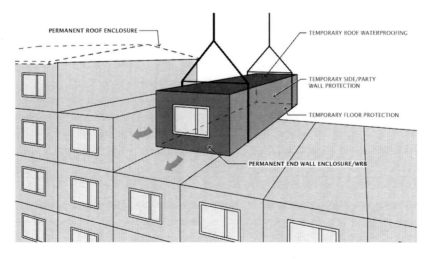

fungal decay[4] or corrosion can be caused by unmanaged moisture throughout the project process. Moisture management and protection is an especially important consideration for modular construction because the transportation and staging present unique circumstances that conventional construction does not typically encounter. Modular buildings also have interstitial spaces between modules where moisture can be trapped unnoticed. Additionally, the connections and tie-ins between the modules require special consideration not seen with conventional construction. If not properly protected, it does not take long for water ingress to lead to fungal growth. Due to these inherent risks, modular buildings require a resilient design that does not allow water ingress to occur prior to the completion of the project.

This paper highlights typical sources of moisture in modular construction, outlines moisture protection measures throughout all stages of the project, provides a summary of recommendations for moisture management, and builds a case for the development of a standard for moisture management in modular construction. The information obtained for this paper results from project design experience, research via forensic investigations, and interviews with housing providers, modular manufacturers, and technical professionals in the modular construction industry.

Sources of Moisture in Modular Construction

Moisture and water ingress can occur at any stage in a module's life cycle from when it leaves the factory to postoccupancy. Typically, moisture damage occurs during storage, transportation, complexing, and postoccupancy. This section outlines potential sources of moisture and water ingress into modules at each of these stages. These typical sources of moisture are based on experiences with past projects and interviews with industry professionals.

STORAGE OF MODULES

Modules are often stored outside the factory before shipping and then staged at the project site prior to being craned into place. This storage and staging in an uncontrolled exterior environment can sometimes be longer than expected due to other project circumstances such as supply-chain delays or inclement weather.

A common source of moisture during storage is the continued wetting of the modules at areas that were not sufficiently protected after manufacturing or areas that were damaged during transportation. Continued damage to the temporary protection can also occur during storage due to birds/rodents or ultraviolet (UV) exposure. If the temporary protection of modules does not incorporate a drainage strategy for the top of the module that acts as a temporary roof, this can lead to ponding water, especially if there are unsupported cavities for shafts and stairwells (**fig. 2**). The ponding of water can lead to water ingress at any areas where the temporary protection has been damaged and may be trapped within the assembled building if not addressed prior to module assembly.

FIG. 2 Ponding water on module roof (A) and condensation at unsupported cavities (B).

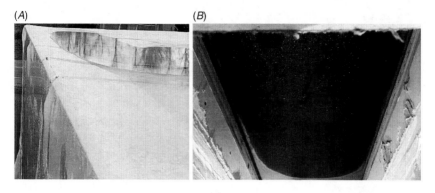

Exterior doorways that are designed to be protected with overhangs when completed or interior corridor doorways can also experience water ingress during storage before complexing. These components typically do not have the same overhang protection in place during storage as when they are assembled in the final building and thus can lead to water ingress. Water ingress can also occur at these locations during transportation (**fig. 3**).

TRANSPORTATION OF MODULES

The transportation of the modules from the factory to the project site can pose a risk of water ingress. The high-speed transportation of the modules can introduce higher forces than what the module will typically experience during its use. For example, if a module is being transported over land during inclement weather, driving rain will be a factor during transportation but may not typically occur when at its final location. The modules may also be transported through different climates and altitudes or even be barged over waterways and oceans.

FIG. 3 Water ingress at doorways.

A common source of moisture in modules is due to damage of inadequate temporary protection during transportation. This can be due to movement of the module on the transport, the wear of the tie-down straps on the protection, wind shear, or potential road debris. Another common source of moisture is at unprotected areas of the module (fig. 4). Sometimes the bottom of the module is left unprotected during transportation, exposing the module materials to moisture. If the bottom is also wrapped in a loose-laid temporary protection such as shrink wrap, water entering from damaged locations can pool below the module and increase exposure of wood sheathing to liquid water.

Finished exterior sections of modules, complete with cladding and fenestration, are often left unprotected. Leaks at unprotected windows, doors, and other exposed components during transportation are common because they are exposed to higher loads during transportation than typical use (fig. 5).

FIG. 4 Typical damage to protection due to tie-down straps (A), wetting of modules at unprotected areas (B), and pooling of water at the underside (C).

FIG. 5 Damage due to water ingress during transportation at unprotected windows.

COMPLEXING OF MODULES

During the complexing of modules, moisture can enter into the interstitial spaces between the modules during placement (**fig. 6**). This can occur if modules are craned during inclement weather or if the tie-in detailing between modules is not completed quickly and inclement weather occurs.

Wet modules complexed before being dried can lead to fungal growth and decay in the interstitial spaces between modules once assembled because there is typically no mechanism to dry out interstitial spaces once complexed (**fig. 7**). Furthermore, leaks can occur at the structural connections between modules when the building has been assembled. These connections often penetrate the waterproofing materials and require appropriate detailing to mitigate water ingress.

POSTOCCUPANCY FLOODING

The final stage in which moisture can damage a modular building is during occupancy. Temporary roofing protection is often left in place on the modules because it simplifies the assembly of the modules. This can have the added benefit of localizing any damage by keeping water on one module if detailed appropriately. However, if flooding does occur, especially small, unnoticed leaks over a period of time, the damage can go unnoticed until it is catastrophic (**fig. 8**). For example, a small

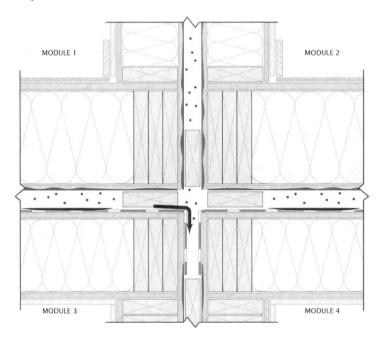

FIG. 6 Moisture trapped in interstitial space between modules with limited ability to dry.

FIG. 7 Fungal growth in interstitial space in between units at walls (A) and floors (B) due to modules being wet before complexed.

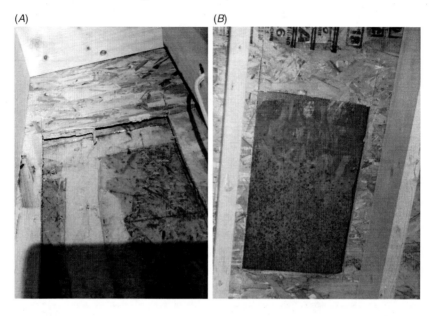

FIG. 8 Severe deterioration of floor joists due to an unnoticed leak from a radiant heating system.

leak can go unnoticed if the moisture is being pooled on the temporary roofing protection. This moisture buildup can remain for a long period before it is noticed, causing moisture damage in the meantime. Even if the temporary roofing is removed before assembly, the double floor/ceiling inherent with modular construction makes leak detection more difficult than with conventional construction. With conventional construction, the ceiling of the lower unit and the floor of the upper unit share ceiling/floor joists and typically have only two layers of sheathing (gypsum on the ceiling and plywood on the floor). If a leak occurs within the upper unit, the leak is generally more noticeable because the leakage pathway is only through one assembly.

Moisture Protection Measures

The modular construction process requires careful thought for moisture management at each stage of the project. This section describes moisture management planning, including effective moisture management considerations for each stage of the construction process. These moisture protection measures are based on experiences with past projects and interviews with industry professionals.

MOISTURE MANAGEMENT PLANNING

Planning for moisture management early in the design phase requires a full understanding of each stage of modular housing construction.[5] Due to the front-loaded design work necessary for modular construction, the choices made in the design stage will affect all other stages of construction. A risk evaluation should be undertaken with the entire construction team to determine all aspects of the project that may contribute to moisture exposure throughout all stages of the building. This should include risks in the manufacturing facility and storage yard, during transportation (including distance traveled, method of transportation, time of year, climates encountered), during staging at the project site (length of time stored in exposed location, time of year, weather events), during complexing of modules (weather, moisture content between modules), construction sealing (material compatibility, detailing robustness, weather), and ongoing moisture management (plumbing systems, fire-suppression systems, end-user habits, humidity sensors). The level of the protection systems for the modules and assembly design should meet the identified risk evaluation.

A moisture management plan for the project can be developed on the basis of the outcomes of the risk-evaluation exercise. The plan should be in writing and accepted by all parties involved, including the design and construction teams. The plan will typically incorporate checklists and procedures to prepare for potential sources of moisture and be included in the project specifications. Hygrothermal analysis of project assemblies should be considered to determine the tolerance of the assembly and to assess the risks of fungal growth if the modules are wetted. Consideration of a contingency to the project budget may be prudent to allow for changes in the moisture management plan to address unforeseen circumstances, such as weather events or construction delays. Throughout each stage of the project it is important to monitor and evaluate the effectiveness of the moisture management plan. It is recommended that moisture management planning and the development of a moisture management plan become a standardized process throughout the modular construction industry.

DESIGN STAGE

The design stage is where important moisture management decisions take place. Decisions regarding module size and layout, material selection, quality control/assurance policies, and transportation methods are made at this stage. Moisture management should begin at the design stage and be incorporated into the design documents and specifications.

Moisture management should be considered when deciding where to place interfaces to minimize difficult tie-in details, to avoid unsupported openings (e.g., stairwells and shafts) during transportation and storage, and to minimize moisture damage from potential flooding postoccupancy.

The air and water control layers must be designed to be continuous once the modules are assembled.[6] This includes considerations for on-site tie-in details, ensuring there is sufficient space between modules to provide quality seals and ensuring compatible materials are used.[7]

Incorporating crawl spaces or basements into the conditioned space during the design stage is recommended. If crawl spaces or basements must be unconditioned, nonmoisture-sensitive materials should be used on the underside of the modules to ensure protection from fungal growth.

Temporary protection can be designed to remain in place for permanent use. For example, the module roof membrane can be incorporated into the air/vapor barrier in conventional roof assemblies. If designed to be left in place, the module roof membranes can also provide some degree of flood protection throughout building occupancy. Permanent protection such as self-adhered sheathing membrane can offer excellent protection during transportation and also double as a compartmentalized air barrier and in situ weather protection for the assembled building.

Site-sealing details are recommended to be included in the architectural drawings (see **fig. 9** for a sample).[5] Isometric details with sequencing can be particularly useful for tricky details. Sequencing of tie-ins is critical to ensure the continuity of

FIG. 9 Sample detail showing partial sequencing of tie-ins.

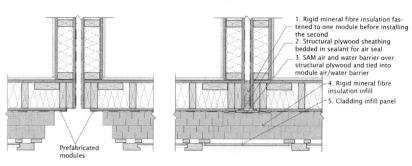

control layers, including water, air, thermal, and often fire. Prestripping of membranes may be necessary, especially at pony walls and foundation tie-ins. Project documents should be clear to ensure proper sequencing occurs for critical barriers and to avoid reverse lapping of membranes. Crane attachment points to the module should be detailed with compatible materials to ensure continuity of critical barriers and be protected from any mechanical damage due to lifting and positioning.

Design strategies should be considered to mitigate any moisture entering into interstitial spaces between modules during construction and occupancy. A resilient design to prevent water from entering interstitial spaces is key, along with strategies to contain, drain, and dry out the spaces if they do encounter moisture. This design may include additional framing for cross-ventilation, redundant internal floor drains located at strategic locations to ensure flooding in adjacent units is minimized, and adding humidity monitoring sensors at strategic locations.

Design strategies should consider the speed of assembly and the ease of integration at module interfaces. For example, the choice of roof assembly is key as a complicated roof structure that may take months to install when the project may encounter inclement weather, whereas the design choice of a simple roof structure could eliminate that risk.

MANUFACTURING AND YARD STORAGE STAGE

The manufacturing stage is an important step in ensuring each module is watertight prior to yard storage and transportation of the modules. Quality control and assurance in the manufacturing stage is critical to ensure each module is properly protected from moisture prior to leaving the manufacturing facility.[8] Water testing, flood testing, and airtightness testing of mock-ups and modules in the manufacturing facility can be particularly useful.[8] Refer to the following for background on potential testing: ASTM D5957, *Standard Guide for Flood Testing Horizontal Waterproofing Installations*,[9] for flood testing; ASTM E1105, *Standard Test Method for Field Determination of Water Penetration of Installed Exterior Windows, Skylights, Doors, and Curtain Walls, by Uniform or Cyclic Static Air Pressure Difference*,[10] for water penetration testing; and ASTM E779, *Standard Test Method for Determining Air Leakage Rate by Fan Pressurization*,[11] and ASTM E1186, *Standard Practices for Air Leakage Site Detection in Building Envelopes and Air Barrier Systems*,[12] for air leakage testing. The inspection of the temporary protection prior to yard storage is another critical part of the manufacturing stage. It is recommended that third-party factory site visits are performed for quality assurance.

It is critical that modules remain dry prior to shipping, especially because many modular manufacturers incorporate moisture-sensitive materials on the interior. To help achieve this, it is recommended that modules not be stored in an uncontrolled environment for extended periods of time without a robust plan for moisture protection. Temporary protection for the modules is critical before the modules leave the factory. There has been a history of success with self-adhered membranes being utilized as the protective membranes for projects in which long storage times or long-distance travel is required (**fig. 10**). The extra cost of a self-adhered membrane is often worth it because moisture damage remediation can be costly and time-consuming, although there is often pressure from clients to use less costly materials for temporary and permanent protection. An advantage of permanent self-adhered or liquid-applied membranes is that if the membrane is damaged in a location, water will only be able to penetrate at that location and not travel between the membrane and the sheathing, therefore minimizing the area of potential damage (**fig. 10**). Although mechanically attached membranes are often used in conventional construction, the risk of damage due to water intrusion is much less compared with modular construction because interior finishes are not yet installed until the building is weathered in. With the added risk of damage, the cost of a better membrane is justified.

An assessment of yard storage conditions and moisture management is critical. Strategies to minimize moisture ingress may include the use of desiccant packs within well-sealed units. If power is available, then consider the use of dehumidifiers or electric heaters, or both. Modules should be stored on raised platforms over dry ground to allow for ventilation and drainage.

If modules are to be stored in the factory yard or on site for more than a day, a drainage strategy should be used to prevent ponding of water. Shrink wrap applied over temporary framing or designing modules to incorporate a sloped roof are potential strategies (see **fig. 11**; temporary framing is typically removed prior to shipping or craning). Care must be taken to avoid damaging existing protective membranes if temporary framing is considered. All horizontal surfaces should be fully supported to prevent temporary roof membranes and shrink wrap from ponding water.

FIG. 10 Advantage of self-adhered membrane versus mechanically attached membrane (arrows indicate water ingress). Note the difference in lateral movement of water.

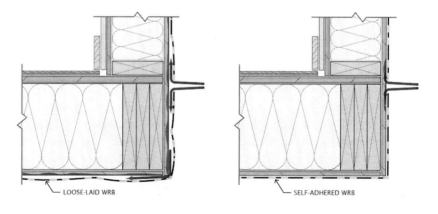

FIG. 11 Loose-laid WRB (shrink wrap) versus self-adhered WRB.

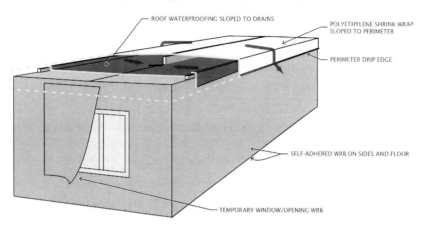

TRANSPORTATION STAGE

There is a high risk of module wetting during the transportation stage of the project. This is especially true if modules are required to be transported over a large distance, over water, through differing climate zones, and during rainy/snowy conditions. Protection of the modules during transportation is essential to ensure that modules are dry when they are craned in place (**fig. 12**).

It is essential that all six sides of the module be protected during transportation.[6] If transportation is over a large distance or includes being barged over water, permanent, self-adhered, or liquid-applied membranes should be considered over nonadhered, temporary protection. Temporary protection should also be considered in

FIG. 12 Temporary membrane protection used for modules transported via barge (A) and truck (B).

addition to the self-adhered or liquid-applied membranes. This extra layer protects the membranes from damage and exposure, including UV exposure during storage and transport. Some membranes have short allowable exposure to UV. Drainage holes at the underside of modules should be provided to prevent water pooling if shrink wrap is used for all six sides. High-speed travel (ground transport) can induce driving rain conditions and cause damage to loose-laid membranes such as mechanically attached housewraps or loose-laid temporary roofing.

Ensure that membrane protection, whether it is temporary or permanent, is protected from abrasion from transportation tie-down straps and other connections. Membranes should also be fully supported at any cavities to avoid sagging and for worker safety when accessing the modules. Road dirt and construction debris such as sawdust should be cleaned from the modules prior to placement because this can lead to accelerated fungal growth if left in place.

Interior moisture management during shipping may also need to be considered for long-duration transport (such as by ships) and transport through different climate zones (such as over cold mountain passes, from warm-humid zone factories to cold climate sites, etc.). Standard shipping desiccants such as calcium chloride can be used to maintain relative humidity levels within well-sealed modules to avoid condensation and decrease the potential for fungal growth (**fig. 13**). Calculations should be

FIG. 13 Desiccant packs within modules during storage and transportation.

undertaken, including airtightness testing and desiccant absorption rates, to ensure the efficacy of the desiccant over the period it is needed.

Alternately, the venting of modules by passive air inlets or other openings may suffice during cool and dry weather and shorter transportation periods. Power supply is a challenge during the transportation stage; therefore, it is recommended not to be reliant on power for heating/dehumidification as part of a drying strategy.

SITE-PREPARATION STAGE

Site preparation is important for ensuring that modules will remain dry and protected from moisture once they are placed on site. Site conditions, foundation preparation, and drainage should be in place to ensure that the foundation, crawl space, or basement is dry and remains dry prior to craning the modules into position. If modules are placed on a concrete podium, drains should be installed to allow drainage during construction if water gets under the modules. After construction, the drains can be capped with a clear cover on the underside of the podium slab. This "port" can be routinely checked for water throughout the life of the building to detect leaks, such as plumbing leaks.

Groundwater and site-specific moisture conditions should be confirmed by qualified personnel. Moisture considerations should include sufficient drainage and active water removal if required. If craning of modules occurs during wet conditions, active drying, such as heating and ventilation, should occur to ensure that moisture does not remain trapped in the space below modules directly above grade.

COMPLEXING STAGE

Construction sealing of each module to the site-built foundation and to each other is a critical stage in modular construction. This prevents any moisture from entering into modules, the interstitial space between modules, and cavities such as crawl spaces before the building is fully sealed. Following the construction sealing steps thoroughly set out in the design stage is important during the complexing stage. If sealing in a wet or cold situation, surfaces may need to be dried or heated, or both, to ensure the material will reliably adhere to the surface. Depending on the climate, it may be important to use materials that can be applied to damp, cold surfaces for sealing in inclement weather.

POSTOCCUPANCY MOISTURE MANAGEMENT

There are major differences in leakage paths in modular buildings versus traditionally constructed buildings. For example, there are multiple layers of sheathing between modules, including double framing in the floors and ceilings of units, and in many cases an impermeable roofing membrane is left in place between each unit that can inhibit leak detection compared with a traditional building. Water may not be visible for an extended period of time and may lead to moisture issues, including fungal growth and decay. Humidity sensors placed in strategic concealed locations such as the interstitial space between modules, the roof system, and crawl spaces, connected to a central monitoring system or alarms, can be beneficial to detecting leaks early so the building can be dried out quickly prior to damage. The use of waterproof flooring and tanking around doorways/thresholds can be warranted, as well as incorporating additional drainage to mitigate in-unit flooding. A maintenance manual for the building is recommended to include key aspects of remediation and protocols for typical flooding events because maintenance staff or emergency-response groups may not be familiar with how modular construction is different from traditional buildings.

Summary of Recommendations

Moisture management and protection is important for modular buildings because modular construction presents unique circumstances that conventional construction does not typically encounter. Key technical recommendations for future projects to help mitigate these risks are summarized in **table 1**. There is currently a lack of standards for the moisture management of modular construction. **Table 1** summarizes the technical recommendations for projects to mitigate the risk of moisture damage throughout the construction process from design stage to postoccupancy. We recommend that these recommendations be considered in the development of a standard for moisture management for modular construction.

Conclusion

Moisture management in modular construction is especially challenging due to unique stages in its life cycle that can increase the potential of water ingress into each module. Common sources of water ingress occur during the storage of modules after manufacturing or on site after transportation, or during transportation itself. Modules that are complexed while wet can experience fungal growth and decay due to inherent building characteristics, such as interstitial spaces, which can lead to a higher risk of moisture accumulation.

Each stage of the modular building process requires careful thought with regard to moisture management. There are many protection measures that can be taken at each project stage to ensure that modules arrive and are complexed without moisture damage. A robust risk analysis and moisture management plan are essential for determining the steps necessary to ensure a successful project. This includes early planning and buy-in from all members of the project team to incorporate moisture management strategies unique to modular construction. Robust protection of the modules during storage, transportation, and complexing is essential for mitigating the risk of moisture ingress at each stage. Due to the concealed interstitial spaces unique to modular construction, a leak detection system is recommended between units to alert building operators of potential in-service leaks. Finally, there is a lack of standards for moisture management for modular construction. The development of a standard or standards could improve the process of delivering dry modules to project sites and improve the quality, consistency, and confidence of modular construction.

TABLE 1 Summary of technical recommendations

Stage	Recommendation
Design stage	1. Perform a risk evaluation for each stage of the project. 2. Require a moisture management plan. 3. Consider hygrothermal analysis of building assemblies to determine the tolerance of the assembly and to assess the risks of fungal growth if the modules are wetted. 4. Design for temporary roof drainage of modules. 5. Ensure sealant material compatibility. 6. Provide thorough sealing details with sequencing.
Manufacturing	1. Ensure quality-assurance tests are performed. 2. Require factory site visits by party responsible for building enclosure. 3. Ensure temporary protection is inspected prior to modules leaving factory. 4. Consider self-adhered membrane protection for long-distance travel and travel over water. 5. Provide drainage holes at underside of modules if shrink wrap is used for all six sides.
Transportation	1. Provide six-sided protection of modules. 2. Ensure membranes are protected from abrasion during travel and craning. 3. Ensure all horizontal surfaces are fully supported.
Site preparation	1. Ensure crawlspaces are dry before placement of modules. 2. Use nonmoisture sensitive materials for unconditioned crawlspaces. 3. Ensure adequate site and foundation drainage.
Complexing	1. Ensure modules are clean and dry prior to complexing. 2. Ensure any moisture from roof of lower modules is removed prior to placement of subsequent modules. 3. Ensure appropriate waterproofing detailing is provided for any craning attachments. 4. Ensure sealing materials are compatible. 5. Ensure proper sequencing is provided.
Postoccupancy	1. Consider placing humidity sensors in interstitial spaces between units. 2. Consider waterproof flooring that ties into the internal drains. 3. Provide a maintenance guide for building managers, firefighters, and remediation teams.

ACKNOWLEDGMENTS

We would like to acknowledge BC Housing for the funding of previous project reports related to modular housing research.[2,5]

References

1. R. E. Smith and T. Rice, *Permanent Modular Construction: Process, Practice, Performance* (Charlottesville, VA: Modular Building Institute, 2015).
2. E. A. Henderson, *Technical Bulletin No. 013: Modular Construction for Energy Efficient, Affordable Housing in Canada* (Vancouver: RDH Building Science Inc., 2019).

3. W. Huss., M. Kaufmann, and K. Merz, *Building in Timber: Room Modules* (Munich: Detail Practice, 2019).
4. P. B. van Niekerk, C. Brischke, and J. Niklewski, "Estimating the Service Life of Timber Structures Concerning Risk and Influence of Fungal Decay—A Review of Existing Theory and Modelling Approaches," *Forests* 12 (2021): 588, https://doi.org/10.3390/f12050588
5. K. Jang, *Modular Housing Moisture Problems and Solutions* (Vancouver: RDH Building Science Inc., 2021).
6. BC Housing, "Appendix D: Modular Construction Methods," *BC Housing Design Guidelines and Construction Standards* (Burnaby, Canada: Author, 2019).
7. BC Housing, *Modular and Prefabricated Housing: Literature Scan of Ideas, Innovations, and Considerations to Improve Affordability, Efficiency, and Quality* (Burnaby, Canada: Author, 2014).
8. T. J. Harrell, J. P. Pinon, and C. D. Shane, "Building Enclosure Design for Modular Construction" (paper presentation, 3rd Residential Building Design and Construction Conference, State College, PA, March 2–3, 2016).
9. *Standard Guide for Flood Testing Horizontal Waterproofing Installations*, ASTM D5957(2013) (West Conshohocken, PA: ASTM International, approved May 1, 2013), https://doi.org/10.1520/D5957-98R13
10. *Standard Test Method for Field Determination of Water Penetration of Installed Exterior Windows, Skylights, Doors, and Curtain Walls, by Uniform or Cyclic Static Air Pressure Difference*, ASTM E1105-15 (West Conshohocken, PA: ASTM International, approved August 1, 2015), https://doi.org/10.1520/E1105-15
11. *Standard Test Method for Determining Air Leakage Rate by Fan Pressurization*, ASTM E779-19 (West Conshohocken, PA: ASTM International, approved January 1, 2019), https://doi.org/10.1520/E0779-19
12. *Standard Practices for Air Leakage Site Detection in Building Envelopes and Air Barrier Systems*, ASTM E1186(2017) (West Conshohocken, PA: ASTM International, approved July 15, 2017), https://doi.org/10.1520/E1186-17

STP 1635, 2022 / available online at www.astm.org / doi: 10.1520/STP163520210015

Xia Cao[1] and Nicholas D. Anderson[1]

Evaluation of Chemical Permeation Performance of Polymeric Below-Grade Waterproofing Membranes

Citation

X. Cao and N. D. Anderson, "Evaluation of Chemical Permeation Performance of Polymeric Below-Grade Waterproofing Membranes," in *Building Science and the Physics of Building Enclosure Performance: 2nd Volume*, ed. D. J. Lemieux and J. Keegan (West Conshohocken, PA: ASTM International, 2022), 269–282. http://doi.org/10.1520/STP163520210015[2]

ABSTRACT

Chlorinated solvents, such as tetrachloroethylene and trichloroethylene, are among the most widespread groundwater and surface-water contaminants in the United States. As the construction industry continues to build larger and deeper buildings into the water table, there is a growing probability that the pre-applied water mitigation systems used in these buildings will be exposed to contaminated groundwater and soils over the course of their service life. Understanding how different membranes and contamination exposure affect the chemical permeation rate is essential to being able to predict the waterproofing's long-term performance as a chemical barrier and ensure indoor air quality is not affected. Using existing standards as a reference, we developed a method to measure the chemical permeation rate of the below-grade waterproofing membranes at different temperature and concentration gradients. This method can accelerate the evaluation process, especially with low-permeable membranes, which typically require longer evaluation times. In addition, this method enables the analyst to build a correlation between chemical permeation rate, temperature, and time. Thus, the method presented

---

Manuscript received January 15, 2021; accepted for publication April 23, 2021.
[1]GCP Applied Technologies, 100 Research Dr., Wilmington, MA 01887, USA N. D. A http://orcid.org/0000-0001-5849-9308
[2]ASTM Second Symposium on *Building Science and the Physics of Building Enclosure Performance* on April 24–25, 2022, and June 12, 2022 in Seattle, WA, USA.

Copyright © 2022 by ASTM International, 100 Barr Harbor Drive, PO Box C700, West Conshohocken, PA 19428-2959.

ASTM International is not responsible, as a body, for the statements and opinions expressed in this paper. ASTM International does not endorse any products represented in this paper.

herein provides a new way to compare the long-term performance of different membranes and can play an important role in the design of water-mitigation systems for buildings.

Keywords

chemical permeation, permeation rate, groundwater, contamination, waterproofing membrane, TCE, trichloroethylene

Introduction

Waterproofing membranes have been widely used to protect the building and its foundation from water and moisture ingress. As the awareness of the quality of occupied spaces increases, indoor air quality is becoming an important factor to consider during the design of a building. In most areas where basements are constructed, the risk of volatile organic compounds (VOCs) and gas migration from the soil and groundwater is normally very low. As a result, waterproofing and mitigation designs are able to manage the VOCs and gas risk.

When buildings are constructed at contaminated sites, near landfills, open mines, or other locations that could produce large volumes of VOCs, there are risks of chemical and vapor intrusion into occupied spaces. It is important to have a barrier that not only stops moisture and water but also reduces or slows down the passage of liquid chemicals and vapors, making the mitigation system more effective in reducing the intrusion of chemicals and vapors. Furthermore, an effective barrier system can reduce the cost of site remediation and expedite construction.

Waterproofing membranes can either be applied as a sheet membrane, a liquid, or a composite system. The system can be installed in two different construction practices: pre-applied or post-applied. Pre-applied waterproofing refers to the system in which waterproofing is applied before concrete placement, whereas post-applied waterproofing is installed after the main structure is formed. In both applications, the ability of the waterproofing system to form an intimate bond to the concrete structure is critical for preventing water migration along the interface and achieving ultimate building protection from water. Typical sheet membranes include a plastic film/sheet and a bonding layer. The plastic sheet is generally considered the primary means of protection against water and other chemicals originating in the soil and groundwater. As the construction industry continues to build larger and deeper buildings, there is an increasing probability that membrane systems used in these buildings will be exposed to contaminated groundwater and soils over the course of their service life.

Understanding the permeation performance of membranes to various volatile organics and how the aging and contamination exposure of the membranes affect the permeation is essential to ensuring the membrane system functions as a long-term building protection. Extensive research in academia and industry have been carried out to evaluate the permeation and diffusion performance of membranes. The method, which closely mimics barrier performance for below-grade applications, is a permeation/diffusion method.[1,2] In these methods, the membrane is sealed between two cells; the source cell is filled with a permeant (pure chemical or solution of known composition and concentration), and the receptor cell is filled with a reference fluid of known composition. With the membrane constantly in contact with the permeant, the overall permeation takes place from the source (higher concentration) to the receptor (low concentration). Solution samples collected from the cells during the test can be analyzed by gas chromatography-mass spectrometry to quantify the contaminant concentrations. The diffusion coefficient can be derived by solving the governing differential equation of mass transfer subject to the appropriate boundary conditions.

Chlorinated solvents, such as trichloroethylene (TCE) and tetrachloroethylene, also known as perchloroethylene (PCE), are among the most widespread groundwater and surface-water contaminants in the United States. In 2011, the U.S. Environmental Protection Agency updated its health assessment for TCE and designated it a Hazardous Air Pollutant commonly entering the atmosphere through volatilization from contaminated groundwater and soils.[3] The focus of this paper is to develop a method to evaluate the permeation and diffusion performance of the membrane with simulated indoor air circulation. TCE was chosen as the contaminant or permeant. A study on how to accelerate the evaluation of a membrane with a low permeation rate and develop a correlation between chemical permeation rate, temperature, and time was also conducted. This method can be a valuable tool for comparing the long-term performance of different membranes and assisting those in waterproofing specification design in choosing the proper product considering the building design and the service life of the system.

Materials and Methods

Although the testing procedures can be applied to many different materials, this study covers three different below-grade waterproofing membranes: Membrane A, a 30-mil (0.762-mm) high-density polyethylene (HDPE) with an adhesive layer and a protective coating layer; Membrane B, a 47-mil (1.194-mm) flexible polyolefin (FPO) with an adhesive layer and a fleece layer; and Membrane C, a 30-mil (0.762-mm) sheet of HDPE/ethylene vinyl acetate (EVA) blend with an adhesive and inorganic granules layer.

Permeation experiments were conducted using ASTM F739, *Standard Test Method for Resistance of Protective Clothing Materials to Permeation by Liquids or Gases under Conditions of Continuous Contact*,[4] Procedure A, Liquid Test Chemical Method, with an open-loop configuration. The open-loop procedure was chosen because it simulates the indoor air circulation well. The liquid test chemical used was reagent-grade (99.5%) TCE from Millipore Sigma or a saturated TCE solution made by mixing the TCE with distilled water. For the latter solution, liquid TCE was kept at the bottom of the storage container to ensure the solution was fully saturated at all times. UHP nitrogen gas (N_2) was used as the collection medium for all testing. Dimensions of the glass permeation cell and parameters of the test apparatus are provided in table 1. A schematic illustration of the test apparatus is shown in figure 1.

Samples were taken by an Agilent 6890N gas chromatograph equipped with a flame ionization detector (GC-FID) outfitted with an automatic gas sampling valve. The 29.60 m × 280 µm × 1.00 µm capillary column used a helium carrier gas at a constant flow rate of 1.3 mL/min. The column temperature was held constant for the length of the test at 120°C. Sampling occurred every 15 min for 30 s and was preheated to 103°C prior to injection into the column. To ensure accurate determination of initial breakthrough time when a very low concentration is anticipated, splitless injection was used to allow the entire gaseous sample into the column without dilution, maximizing the detection of the analyte. To ensure no residual contamination between specimen, the gas valve was purged to vent with helium prior to every sampling.

TABLE 1 Characteristics of the glass permeation cell

Parameter	Value
Volume of collection medium cell	100 cm³
Volume of liquid test chemical cell	30 cm³
Area of exposed sample material	18.79 cm²

FIG. 1 Experimental setup of open-loop permeation system with calibration addition.

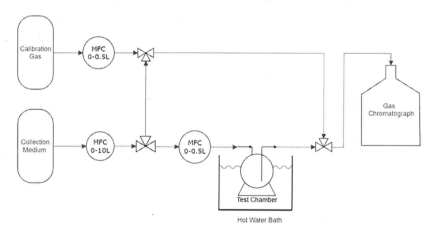

A calibration curve for TCE in the carrier gas (N_2) was created using a known standard of 497 ppm TCE in N_2. This standard was diluted in-line with additional N_2 and was injected into the 15:1 split GC-FID. By varying the flow ratios of the TCE standard and additional N_2, a range of TCE concentration was achieved. **Figure 2** shows the calibration curve to determine the TCE concentration in N_2 that can be used to calculate the permeation rate. The setup is able to achieve detection sensitivity lower than the 0.1 µg/cm^2/min required for open-loop testing per ASTM F739.[4]

In all testing, the waterproofing membrane was oriented such that the face in contact with groundwater during normal commercial usage is in contact with the TCE solution. Prior to each test, the permeation cell was cleaned in an ultrasonic bath at 60°C to remove all contaminants. The assembled cell was connected via PTFE tubing to the gas sampling ports and fully immersed in a circulating water bath covered with hollow polypropylene balls to minimize temperature fluctuation and water evaporation. Testing occurred at various temperatures (25, 40, 50, and 60°C; all ± 0.1°C). Acting as the collection medium, gaseous nitrogen flowed at 100 mL/min across the membrane, with a sample analyzed every 15 min. Each test continued until a steady-state permeation rate could be achieved.

In this study, the normalized breakthrough detection time and steady-state permeation rate were the key parameters for evaluating the overall chemical protection different materials can provide. Both values, as an example, can be seen in **figure 3** and **figure 4**, respectively. The steady-state permeation rate was achieved once the measured permeation rate remained constant over several measurements, as shown

FIG. 2 Calibration curve of TCE in nitrogen.

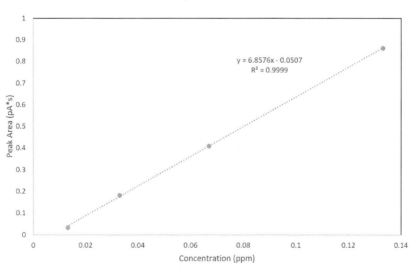

FIG. 3 Steady-state permeation rate.

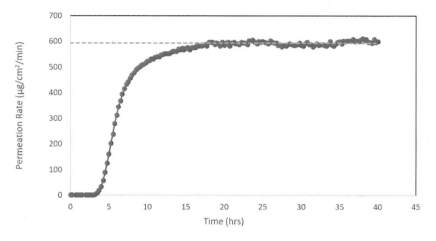

FIG. 4 Normalized breakthrough time.

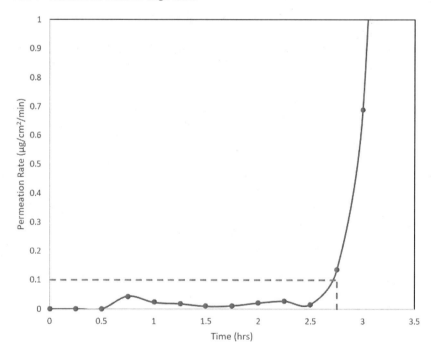

by the dashed line in **figure 3**. When looking closely at the origin of the same data set, **figure 4** shows the point at which the permeation rate reached 0.1 µg/cm^2/min, the normalized breakthrough time as defined in ASTM F739.[4]

Results and Discussion

The physical process of chemical permeation involves the diffusion of permeant through a membrane from high concentration to low concentration in three steps: adsorption, diffusion, and desorption.[2,5] When these three steps have reached equilibrium, the overall permeation rate is dependent on the permeant's concentration gradient across the membrane, the membrane's intrinsic permeation coefficient, temperature, and pressure. For single-layer membranes, there are experimental techniques and theoretical models[6] for characterizing the parameters of permeation performance. However, most waterproofing membranes are composites with multiple layers that make the theoretical modeling difficult.[6] Therefore, in this study, we focused on the overall permeation rate of the samples. This total rate is also useful in real-world applications because building designers mostly care about overall membrane performance, not individual layers.

Most diffusion studies[2,5] are performed at room temperature, which, simply put, cannot account for the variety of environmental conditions and in which chemical permeation can occur. It is important to note that because the temperature and concentration gradient play a critical role in the chemical permeation through the membrane, a good understanding of membrane product performance through experimental evaluation and empirical modeling at various temperature and contamination ranges is essential for choosing the right product, especially for high-risk, sensitive, and challenging applications.

REAGENT-GRADE TCE AS SCREENING TOOL

The open-loop setup with undiluted reagent-grade TCE has the highest possible concentration gradient. Although this concentration gradient is not likely to exist in real-life commercial scenarios, it is a useful evaluation because the permeation rate can quickly reach a steady state and is easier for instruments to detect because it represents an "upper bound." The setup with reagent-grade TCE can be used as a quick initial filter or concept check prior to evaluation at boundary conditions closer to reality. The screening also filters out membranes that have poor chemical resistance to the permeant.

Figure 5 and **table 2** summarize the permeation-rate change of TCE through Membrane A at different temperatures over time. At an ambient temperature of 25°C, it took 10.5 h to detect TCE in the receptor cell. The transition time from initial detection to steady-state permeation at 182 $\mu g/cm^2/min$ was close to 30 h. At 40°C, TCE could be detected after 2.75 h, and the transition to a steady state around 593 $\mu g/cm^2/min$ happened close to 10 h. When the temperature was at 50°C and above, the permeation could be detected at around 15 min, and the steady-state permeation rate was higher due to the increased molecular mobility and partition rate at the higher temperatures.

A parallel comparison of different membranes at 60°C can be a quick screening tool for choosing the better option to cover the broad temperature range.

FIG. 5 TCE permeation rate through Membrane A at different temperatures (reagent-grade TCE).

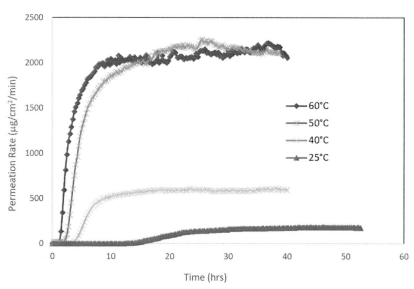

TABLE 2 Normalized breakthrough detection time and steady-state permeation rate (reagent-grade TCE)

	Membrane A			
Temperature	25°C	40°C	50°C	60°C
Normalized breakthrough detection time (h)	10.5	2.75	0.25	0.25
Steady-state permeation rate (µg/cm²/min)	181.9	592.8	2,145.2	2,119.1

Membrane B was unable to withstand the chemical attack of the reagent-grade TCE. **Figure 6** demonstrates how Membrane B performed during the evaluation at 60°C. Membrane B showed no permeation for a short time period as the chemical began to penetrate the barrier. Soon after, the main waterproofing layer started to soften and dissolve (Point a), resulting in a sharp increase in the permeation rate. The softened polymer started to fill gaps between the fleece bonding layer (Point b) before further breakdown (Point c), filling and blocking the receptor cell's outlet (Point d). Membrane C was unable to achieve a tight seal in the test cell at the beginning of the evaluation with reagent-grade TCE at 60°C; thus, no data could be collected.

The comparison between Membrane A and Membrane B (prior to its breakdown in TCE) in **figure 7** demonstrates better barrier properties of Membrane A

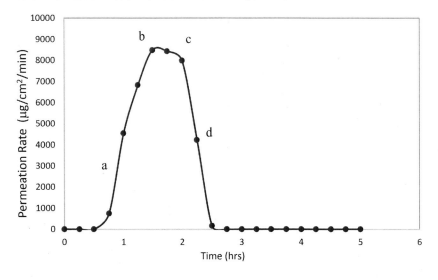

FIG. 6 Permeation rate of Membrane B at 60°C against TCE.

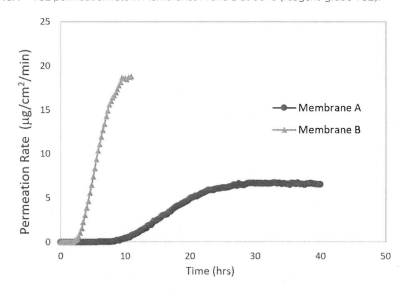

FIG. 7 TCE permeation rate in Membranes A and B at 60°C (reagent-grade TCE).

over Membrane B at 60°C. Membrane B had more than four times the permeation rate prior to membrane breakdown and had an earlier initial detection compared with Membrane A. Unlike Membrane B, Membrane A was visually undamaged and remained a waterproof barrier throughout the exposure to the reagent-grade TCE during the evaluation period.

SATURATED TCE TO MIMIC REALISTIC ENVIRONMETAL CONDITIONS

TCE concentrations are normally low in surface water but can be higher in groundwater systems due to limited volatilization and biodegradation. From one reported study of 481 municipal/communal and 215 private/domestic groundwater supplies, contaminated groundwater samples contained TCE at average maximum concentrations between 25 and 1,680 µg/L.[7] Concentrations may be even higher if contamination has occurred locally and leaching has taken place.

To provide a valuable evaluation and cover the worst-case scenario, a saturated concentration of 1.1 g/L (1.1×10^6 µg/L)[8] was chosen for the additional investigation. The results are summarized in **tables 3–5** and **figure 8**.

When Membrane A was evaluated with saturated TCE at 40°C (**Fig. 8A**), TCE could be detected at 8.5 h and took around 30 h to reach a steady-state permeation rate of 6.7 µg/cm^2/min. For the reagent-grade TCE at the same temperature, breakthrough detection with Membrane A occurred at around 2.75 h and transitioned to

TABLE 3 Peak permeation rate and normalized breakthrough detection time of saturated TCE solution through Membrane A at different temperatures

	Temperature (°C)			
	25	40	50	60
Steady-state permeation rate (µg/cm^2/min)	N/A	6.7	9.6	20.3
Normalized breakthrough detection time (h)	No detection; stopped at 21.75 h	8.5	4.5	2.25

TABLE 4 Peak permeation rate and normalized breakthrough detection time of saturated TCE solution through Membrane B at different temperatures

	Temperature (°C)			
	25	40	50	60
Steady-state permeation rate (µg/cm^2/min)	16.8	19.2	26.1	50.8
Normalized breakthrough detection time (h)	0.25	2.25	1.25	0.75

TABLE 5 Peak permeation rate and normalized breakthrough detection time of saturated TCE solution through different membranes at 60°C

	Membrane		
	A	B	C
Steady-state permeation rate (µg/cm^2/min)	20.3	50.8	22.48
Normalized breakthrough detection time (h)	2.25	0.75	1.25

FIG. 8 TCE permeation rate of saturated TCE through (A) Membrane A at different temperatures, (B) Membrane B at different temperatures, and (C) through different membranes at 60°C.

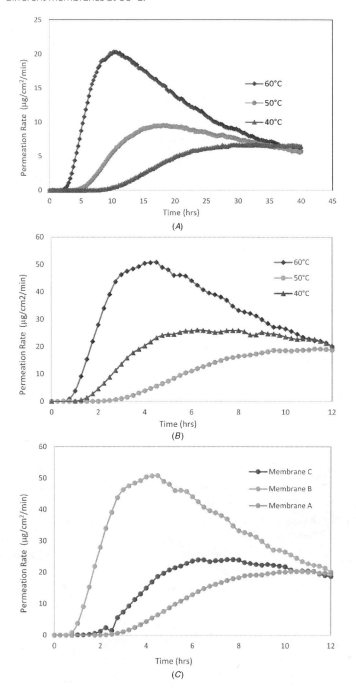

a steady-state permeation rate of 593 µg/cm²/min around 10 h (**Fig. 5**). TCE took three to four times longer to reach a detectable permeation rate with saturated concentration testing, and the steady-state permeation rate was only 1.1% of the rate with reagent-grade TCE. These differences are close to the permeation rate extrapolated from reagent-grade TCE based on Fick's first law.[2] It also indicates the similarity of physical properties of the membrane in the pure and saturated TCE solution for Membrane A at 40°C:

$$f = -D_g \frac{dC_g}{dz} \quad (1)$$

where:

f = permeation rate, unit time, and area,
D_g = permeation coefficient,
C_g = concentration of contaminant,
z = distance parallel to the direction of transport (or thickness of the membrane in this case), and
dC_g/dz = concentration gradient across the membrane.

When the temperature was raised to 50°C and above, the permeation rate through Membrane A increased, and the time to breakthrough happened earlier. Both permeation profiles at 50°C and 60°C showed permeation-rate declines after the peak. The same was observed for Membrane B, with a higher permeation rate at all temperature conditions. The transition point at which the rate began to decline occurred after approximately 15% of the TCE, initially in the source cell, had permeated through the membrane (**fig. 9**). This was consistent across all tested temperatures. The decline resulted from the permeant TCE concentration reduction in the source cell over time. Unlike undiluted TCE, which can maintain the consistent high concentration, the concentration reduction in the source cell reduces the concentration gradient, and therefore the permeation rate, over time.

To verify the hypothesis, the source cell was drained and replenished with a fresh volume of saturated TCE solution during 60°C testing. **Figure 10** shows the almost immediate increase in permeation rate due to the rise of concentration gradient. To avoid concentration reduction in the source cell for the open-loop evaluation, constantly replenishing or circulating of saturated solution is important, especially for the evaluation at a high temperature with a raised permeation rate.

Compared with Membrane A in a saturated TCE solution, Membrane B had a higher permeation rate and earlier detection and reached a peak permeation rate faster across all temperatures (**Fig. 8A** and ***B***). The overall trend was similar to reagent-grade TCE testing. For comparisons between multiple options, high temperature and high concentration can be a quick screening tool. However, the permeation rate at steady state may be a magnitude higher compared with the actual application condition depending on the temperature and concentration used for screening. **Figure 8C** shows the differences between various membranes at 60°C. The HDPE-based membrane recorded the lowest permeation rate against TCE, followed

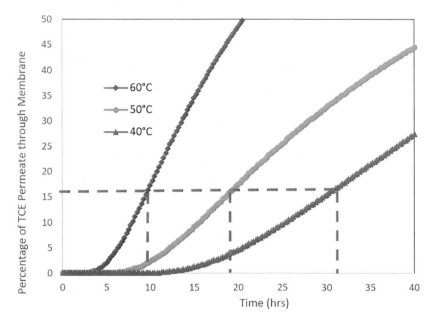

FIG. 9 Percentage of TCE in the source cell that permeated through Membrane A (saturated TCE).

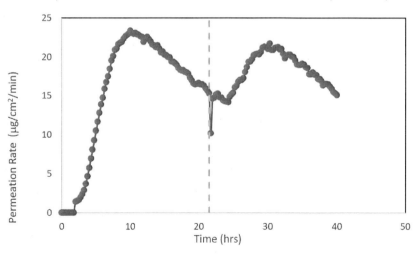

FIG. 10 Replenishment of source cell with saturated TCE at mid-test (dashed line).

by the HDPE/EVA-based Membrane C. Although Membrane B was the thickest product, the FPO-based membrane had the highest permeation rate out of the membranes tested. For gas barriers designed to reduce air-contaminant diffusion from the ground, membranes with lower permeation rates are desired.

Summary

This study was aimed at developing a methodology to evaluate the TCE permeation performance of waterproofing membranes using an open-loop approach that considers the indoor air exchange in an occupied space. With ASTM F739 as a reference, we investigated the permeation of TCE through various waterproofing membranes at different temperatures and different concentration gradients. Preliminary findings demonstrated evaluations at high temperature or high concentration gradients can be used as a screening tool for comparing different products and speeding up the evaluation.

References

1. H. August and R. Taztky, "Permeability of Commercial Available Polymeric Liners for Hazardous Landfill Leachate Organic Constituents," in *International Conference on Geomembrane* (St. Paul, MN: Industrial Fabrics Association International, 1984), 151–156.
2. H. P. Sangam and R. K. Rowe, "Migration of Dilute Aqueous Organic Pollutants through HDPE Geomembranes," *Geotextiles and Geomembranes* 19 (2001): 329–357.
3. U.S. Environmental Protection Agency, "Toxicological Review of Trichloroethylene," http://web.archive.org/web/20200523163244/https://cfpub.epa.gov/ncea/iris/iris_documents/documents/toxreviews/0199tr/0199tr.pdf
4. *Standard Test Method for Resistance of Protective Clothing Materials to Permeation by Liquids or Gases under Conditions of Continuous Contact*, ASTM F739-99a (West Conshohocken, PA: ASTM International, approved June 10, 1999), https://doi.org/10.1520/F0739-99A
5. J. K. Park and M. Nibras, "Mass Flux of Organic Chemicals through Polyethylene Geomembranes," *Water Environment Research* 65 (1993): 227–237.
6. A. Alentiev and Y. Tampolskii, "Prediction of Gas Permeation Parameters of Polymers," in *Material Science of Membranes for Gas and Vapor Separation*, ed. Y. Yampolskii, I. Pinnau, and B. D. Freeman (New York: John Wiley & Sons, 2006), 211–229.
7. World Health Organization, "Trichloroethylene in Drinking-water," WHO, 2005, http://web.archive.org/web/20200919073212/https://www.who.int/water_sanitation_health/dwq/chemicals/trichloroethenemay05.pdf
8. E. W. McGovern, "Chlorohydrocarbon Solvents," in *Industrial & Engineering Chemistry*, (1943), 1230-1239.

STP 1635, 2022 / available online at www.astm.org / doi: 10.1520/STP163520210035

Michael Carter[1]

Evaluation Methods, Hazard Risk Assessment and Mitigation of Building Enclosure Design Elements that are Prone to Ice and Snow Collection and Release

Citation

M. Carter, "Evaluation Methods, Hazard Risk Assessment and Mitigation of Building Enclosure Design Elements that are Prone to Ice and Snow Collection and Release," in *Building Science and the Physics of Building Enclosure Performance: 2nd Volume*, ed. D. J. Lemieux and J. Keegan (West Conshohocken, PA: ASTM International, 2022), 283–297. http://doi.org/10.1520/STP163520210035[2]

ABSTRACT

The collection of ice and snow on a building enclosure can have multiple impacts on design that require consideration. This includes the design of a building's structural snow and ice load capacity, accessibility or egress due to drifting snow, infiltration of snow into openings or accesses, as well as the potential for blockage of mechanical intakes and exhausts by snow. In addition, a key safety consideration is the potential for ice and snow to accumulate and release from a building enclosure, creating a potential hazard to people or damage to property. To address these impacts, microclimate consultants have developed methods of evaluating architectural designs through the analysis of historical meteorological data and site surroundings, in combination with a review of the proposed architectural design details and their exposures. However, given the limited existence of published industry research, guidelines, standards, and building codes on the topic, these types of assessments remain highly experience-based and therefore rely on the knowledge, expertise, and skill of the evaluator to identify

---

Manuscript received March 29, 2021; accepted for publication August 16, 2021.
[1]Microclimate Ice & Snow Inc., 25 Lake Ave. East, Cambridge, ON N3C 2V4, Canada
[2]ASTM Second Symposium on *Building Science and the Physics of Building Enclosure Performance* on April 24–25, 2022, and June 12, 2022 in Seattle, WA, USA.

Copyright © 2022 by ASTM International, 100 Barr Harbor Drive, PO Box C700, West Conshohocken, PA 19428-2959.

ASTM International is not responsible, as a body, for the statements and opinions expressed in this paper. ASTM International does not endorse any products represented in this paper.

potential design impacts and effectively communicate risk, as well as the relative effectiveness of mitigation and management strategies. Addressing these points, a discussion of current and past practices, evaluation methods, problem and risk identification, as well as mitigation strategies will be presented, along with a case study reviewing a common architectural facade element that is prone to ice and snow collection and release.

Keywords
ice, snow, hazard, falling, sliding, release, microclimate, facade, snow loads, barrier, snow guard, ledge, winter, operations, evaluation, risk

Introduction

The collection of ice and snow on buildings surfaces is often uneventful and even pleasing to observe. However, ice and snow accumulation can affect multiple aspects of building design and operation. These include structural loading,[1,2] grade-level egress, building accessibility, service openings, views through skylights, as well as safety due to the potential for hazard to people or damage to property from falling, sliding, or wind-released ice and snow.

Focusing on the latter, this paper examines the evaluation methods of ice and snow microclimate assessment and includes a case study reviewing a common architectural design element that is described as a protruding ledge, concluding with a discussion of recommendations for the building industry.

It is important to highlight that an ice and snow microclimate assessment using the evaluation methods described herein cannot identify or eliminate all risk of a potential hazard or incident occurring[3]; therefore, it is necessary for all buildings located in climates that receive winter precipitation (i.e., snow, sleet, freezing rain, freezing mist and hail) be actively managed and monitored using established operational procedures that prescribe the actions necessary to maintain safe operation of a building during winter conditions, protecting people and property from falling, sliding, or wind-released ice and snow.[4]

Background: Ice and Snow Microclimate Assessment

Microclimate consulting specific to the review for hazardous release of ice and snow from buildings originated primarily out of necessity when building owners and designers would become aware of incidents or potential risks on existing buildings or through past reported events of similar design geometries during design.

A specific highly publicized example of an existing building incident occurred on February 28, 1994, in Chicago, Illinois. This incident resulted in a death and was thought to be the first significant legal settlement of $4.5 million, thereby developing awareness of the issue within the broader industry. At the time it was reported that the man was hit on the head with "a piece of ice described by one police officer" as "the size of a microwave."[5]

As a result of increased awareness within the design and building ownership industries, when a risk was identified by a building owner or designer they would inquire with microclimate consultants with specific knowledge of historical winter weather analysis and ice and snow behavior, to resolve and mitigate identified issues. Subsequently, over time as consultant knowledge and experience grew through the completion of incident and project investigations, it became evident that the number of potential issues and their potential for mitigation impacted a broad range of design topics and disciplines. This ranged from initial design decisions during schematic design all the way through to construction documents and installation, including decisions specific to facade detailing, thermal performance, material surface characteristics, and the incorporation of mitigation.

In recent years it has further been identified that problematic occurrences of ice and snow collection and release are closely linked to advancements in design complexity and envelope performance. These are primarily driven by advancements in design and material technologies, as well as regulatory requirements for energy conservation.[6] Another key influencing factor is the increase in winter storm variability and the severity of extreme events that have been linked to a warming global climate.[7] As a result, these factors are thought to be contributing to the increased reports of icing events on recently completed buildings. One example of the increase in winter storm severity is revealed by a review of long-term historical data summary for Central Park in New York City (1869–2021) produced by the National Weather Service, which focused on snowstorms of 12 in. or greater.[8] Upon review of this summary, it is evident that approximately 50% of the recorded events occurred in the most recent 30 years of the 150-year record. This example, however, should not be construed to conclude that 12 in. of snowfall is required to produce a falling ice and snow event, as an event can occur with significantly less snowfall or freezing precipitation. Instead, the frequency of a problematic event should be thought as an issue that is propagated by or more closely linked to a susceptibility in a building's design features or details combined with exposure than storm severity (i.e., a very susceptible feature or details could perform negatively in all types and magnitudes of winter storms, whereas others may be susceptible only to more severe or unique events such as an ice storm, freezing fog, or a high-elevation in-cloud icing event).

Evaluation Methods for Ice and Snow Microclimate Assessment

MICROCLIMATE DESIGN REVIEW

A microclimate design review is the foundation of any ice and snow evaluation method and is applicable to existing buildings, buildings during the design stage,

and buildings under construction. A design review typically comprises the following basic elements:
- A statistical analysis of historical long-term meteorological conditions local to the site (e.g., number of days annually with 1, 3, 5, and 10 in. of snowfall, various time frames of annual snowfall averages, 24-h extreme frequency analysis, freezing rain or ice storm frequency, etc.)
- An experience-based analysis of the winter microclimate wind flows, based on the statistical analysis results, predicted to influence the accumulation of ice and snow on an existing or proposed building design
- A review of the building design geometry, orientation, thermal properties and architectural details in the context of the predicted winter microclimate and site conditions, identifying issues on the basis of past projects or field or testing experience that can lead to a potential safety hazard or property damage

Once complete, the design review forms the foundation of the more encompassing ice and snow microclimate assessment, which includes incident or design investigation, risk assessment, mitigation development, feasibility, and validation of mitigation.

INCIDENT OR DESIGN INVESTIGATION

An incident investigation is performed when an existing building has experienced a potentially hazardous or damaging ice and snow incident,[9] whereas a design investigation is performed when a risk or vulnerability is identified by a design team member or ownership on the basis of past experience or concern. In these situations a microclimate design review is supplemented with additional available incident information, a review of meteorological conditions at the time of the incident, as well as interviews with observers, building operators, or designers, then progresses to a risk-assessment phase.

RISK ASSESSMENT AND DETERMINATION OF EVALUATION CRITERIA

The evaluation of risk has many variables that affect the final decision regarding the acceptable level of mitigation response, either at the design stage or as a retrofit to an existing building. To assess risk and determine an acceptable response, an evaluation of the current risk situation is completed in combination with a determination of feasible mitigation options that can be presented for evaluation by a client. The assessment of risk includes determining the following:
- Type of problematic condition (e.g., sliding, overhanging or wind-released snow, falling icicles, ice damming, sliding ice sheets, etc.)
- Surface area and repetitiveness of the problem condition on the facade (e.g., conditions repeated at every facade joint or floor plate, etc.)
- Estimated frequency of problematic, severe, or hazardous formations (i.e., the anticipated sensitivity of the condition to snow and ice accumulation and release)

- Sensitivity or requirements of the affected fall zone (i.e., whether ice and snow will fall in an area that requires a high standard of care such as a hospital entrance)
- Availability of operational means of mitigation and temporary restricted access (e.g., whether temporary restricted access for a 2- or 3-day time period once or twice a winter season is acceptable or whether a 2- or 3-day time period once every 3 to 5 years is more appropriate)

Note that the above list is not all-inclusive and can contain many more variables, both common and uncommon, that require consideration.

In performing a risk assessment it is important to consider the following:
- An assessment of this nature does not replace or eliminate the requirements for a separate structural snow load study to address building code requirements and safety concerns.
- An assessment of falling, sliding, or wind-released ice and snow is related to the day-to-day operations of a building (i.e., not correlated to the more extreme events associated with the determination of snow loads) and focused on reducing hazard potential, as well as the acceptability for temporary restricted access or the deployment of warning signs.
- Sliding ice and snow on a building roof or large surface has significant implications for building designers and structural engineers, potentially resulting in unbalanced loading, noise and vibration or impact loads that can damage any obstructions or property along the slide trajectory (i.e., on a building surface, a lower roof, or on the ground below).
- An assessment of this nature cannot eliminate all ice and snow concerns; therefore, site operations will still need to be relied upon to monitor and maintain a safe environment during and after winter storm events.
- In the absence of building codes and standards,[10] it is the responsibility of the design team and / or ownership to determine the acceptable level of mitigation response for the project, including the acceptable level of reliance on building operations required to monitor and maintain safety.

At this point in the process it is important to reiterate and highlight that which was stated above: the absence of established building codes, standards, or guidelines within the industry necessitates the need for building designers, owners, and operators to establish and communicate their risk tolerance and therefore evaluation criteria to determine the required magnitude of mitigation response. This is hopefully based on guidance from a knowledgeable and experienced consultant or project team member. Once established, an evaluation criteria and mitigation response strategy guides the decision-making process for the various identified problem areas of a building.

To further demonstrate the importance of taking the time to discuss and establish project specific evaluation criteria, a news article referencing a lawsuit filed by a bike rental company located near a recently completed supertall building in New

York City is cited.[11] In the article, the rental company claims that the occurrence of a 2019 incident, when ice and snow fell from the construction site and construction equipment of a building under construction, caused a loss of "six days of business and revenue due to the falling ice, and is seeking at least $150,000 plus fees, in restitution."[11] The article quotes the plaintiff in the lawsuit filing that states: "Negligent ice maintenance and removal related to the project at the Central Park Tower caused a shutdown of West 58th Street and several adjoining streets,"[11] and "Current and potential customers were either fearful of or not permitted to walk toward [our] store because their lives were in danger."[11] The plaintiff's lawyers are then quoted as saying "While the city thrives on new development, building owners and contractors have an obligation to ensure their construction projects do not present a danger to the public,"[11] and "Despite previous incidents, warnings and shutdowns, the defendants did not take the necessary steps to ensure public safety."[11]

Even though this action is specific to a building under construction, it offers a key insight into an underlying issue of risk with respect to lost revenue and damages related to the temporary restricting of access due to falling ice and snow. This is emphasized by the fact that the plaintiff is not seeking restitution for physical injury or damage but for a loss of business revenue because their establishment resides in a fall zone from a tall building that required building owners and city officials to restrict access for extended periods of time. Furthermore, it is important to consider the fact that temporary restricted access is often deployed on the basis of a perceived potential for hazardous falling ice, regardless of the magnitude of ice that exists on the building. Therefore, as modern building designs are becoming more susceptible to icing on their exterior surfaces,[6] it is theorized that the instances of temporary restricted access are likely to increase dramatically as more modern buildings are completed, specifically as energy-efficiency regulations implemented by government authorities advance, resulting in greater heat conservation, thus creating colder and more ice-prone building facades.

This raises a question from an evaluation-criteria perspective. Is it sufficient to only identify and mitigate the potential for hazardous ice and snow release from a building, or should any magnitude or instance of possible ice and snow accumulation be identified and evaluated, including the resultant level of mitigation response that would be required to minimize the need for restricted access below? Asking the question in another way, should an ice and snow microclimate assessment ultimately reduce the anticipated number of days or hours that restricted access would be necessary, irrespective of the size or magnitude of released ice or snow that could release, rather than focusing solely on the mitigation of identified hazardous snow or ice formations? To accomplish this higher standard, a more thorough understanding of all potential icing conditions and scenarios would be required by evaluators, as well as a willingness of building designers and owners to accommodate the added effort required to obtain this understanding. This added effort could include:

- Additional detailed computer or analytical analysis and investigation (e.g., a more detailed meteorological analysis of historical weather data or modeled

future data; three-dimensional [3D] computer modeling of wall or roof thermal properties, including the interaction with specific winter weather events; or 3D computer modeling of microclimate wind and snow deposition on building surfaces [see the "Mitigation Validation and Refinement" section below])
- Performance mock-up testing (i.e., a full-scale thermally correct mock-up of the detail in question placed within a climatic chamber or climatic wind tunnel and then subjected to a program of replicated winter precipitation events to investigate winter performance or validate mitigation strategies)
- Acceptance of a more effective mitigation response

MITIGATION DEVELOPMENT, FEASIBILITY, AND VALIDATION
Mitigation Development

The development of an effective and acceptable mitigation strategy for an identified problem condition on a building is often challenging. It requires not only a thorough understanding of the relative effectiveness of various available mitigation strategies in all types of winter weather but also knowledge of implementation for the geometries in question as well as any remaining risks specific to the chosen evaluation criteria. To initiate the process, available mitigation methods need to be evaluated for feasibility and validation potential in the context of the problem condition and then detailed for the individual application. Four categories of mitigation are discussed in the following as examples:

1. Commercially available ice and snow retention and heating cable products are often used as a means of mitigating ice and snow-related issues. However, the application of these products often requires additional expert knowledge to address specific performance criteria for large and complex roof and facade geometries. It is further important to state that the ice and snow retention and heating industries also do not benefit from published performance standards, resulting in a "buyer beware" status that requires due diligence when choosing, specifying, and installing these products.[12] However, some product specific evaluation criteria do exist.[13] To assist with this situation, members of the snow retention industry as part of ASTM Subcommittee E06.55, Performance of Building Enclosures, are in the process of developing testing standards for the snow-retention industry.

2. A second effective method of mitigating hazardous ice and snow release is through architectural design modification. In the simplest form this can include the modification of site usage in the area below a problematic architectural detail, the elimination of a detail from the design, or the addition of a canopy to protect sensitive areas below an architectural detail. However, from a building designer's or owner's perspective these options are often not desirable or practical. As a result, the next level of consideration is the customization or modification of a design detail to reduce risk, while minimizing any impact on design aesthetics and retrofit costs. This method,

however, is a highly experience-based effort that is accomplished through the incorporation of modifications or additions to the architectural design geometry, material surfaces, thermal properties, or a combination thereof.

3. A third method is to melt ice and snow on contact through the addition of an integrated or customized heat system. This type of mitigation method can take the form of conductive, radiant, and isothermal melting systems and typically requires a combination of mechanical engineering input, detailed design parameters (i.e., placement, quantity, and density of heat) and validation testing. However, if heat is being applied to an architectural detail, careful consideration of meltwater is critical to avoid issues of damage and hazard due to ice and icicles produced by the refreezing of meltwater on other parts of the building or on walking surfaces below. Furthermore, the application of heat has implications on design, construction, and long-term maintenance and operational costs. As such, these types of methods are typically used only if the surfaces in question have a large total area that would justify the added design complexity, components, and corresponding costs.

4. A fourth less established method is the application of anti-icing or icephobic coating technologies. These types of coatings are progressing quickly in terms of their development and application; thus, it is hopeful that in the near future they will be sufficiently effective and durable for large-scale applications. However, at this time, these types of coatings have been used only in a limited number of field applications on buildings. In addition, the various investigated technologies have been developed for specific applications and therefore have strengths and weaknesses with respect to the large variability in building materials and winter precipitation characteristics.[14] For example, dry snowfall, wet snowfall, freezing rain, freezing mist, freezing fog, and in-cloud icing all have different collection characteristics on surfaces that are challenging to resolve with a single technological approach.

Whatever method of mitigation is chosen, it is important to consider the variability of winter conditions over time because ice and snow on a building surface can transition and transform, as well as the need for validation and refinement of any mitigation method or strategy through experience-based or evidentiary means of evaluation, or both, including if necessary the specification and completion of performance mock-up testing conducted in a qualified environmental test laboratory.

Mitigation Feasibility

Determining the feasibility of an effective mitigation method depends on many influencing factors, including:
- Design team/ownership interpretation of risk exposure
- Effectiveness of the mitigation
- Constructability of the mitigation
- Cost of the mitigation
- Implication to the architectural design intent

Of these factors the most important is the effective communication to the design team/ownership of the anticipated risk exposure based on the results of the risk assessment and allowing for the corresponding level of mitigation response to be accepted by building designers and owners, balancing risks, costs, and aesthetics.

Mitigation Validation and Refinement

The validation or refinement of mitigation is recommended when experienced-based assumptions or past project experience are not sufficient or directly applicable. Therefore, to increase confidence in the performance of mitigation or to investigate the detailing of mitigation within a specific application, further investigation is recommended. This can be accomplished through one or a combination of the following means:

- *Detailed meteorological analysis*: A detailed meteorological analysis utilizes historical meteorological observational data or data derived from meteorological modeling simulations to determine the reoccurrence frequency and severity of specific meteorological conditions, allowing for a more detailed understanding of snow accumulation or icing on a building and therefore the risk potential and performance requirements for mitigation. As a result, this type of analysis is often used to inform the other types of investigations listed below, as well as the overall risk assessment.
- *Performance mock-up tests*: Performance mock-up tests typically consist of a physical full-scale thermally correct mock-up constructed and configured to replicate a representative portion of the identified architectural detail within the context of the building design and exterior exposure. This includes any surrounding or supporting roof or facade elements that could influence the test results. The purpose of the test is to expose the mock-up to a wide range of typical and severe winter storm event conditions that will be experienced by the completed building. The tests are performed in a climatic wind tunnel or a large climatic chamber providing the ability to perform controlled and repeatable test plans.
- *Sample testing*: Testing can also take the form of individual sample tests that are specific to the performance of individual facade material surfaces. This type of testing is typically performed on the basis of established test methods and procedures but can also consist of observational-based qualitative testing designed to assess the behavior of ice and snow on a surface during the collection and melting process.
- *Computer modeling simulation*: Computer modeling simulations can provide a more thorough and detailed understanding of ice and snow deposition patterns on building surfaces. For example:
    1. Finite area element (FAE) models combine wind tunnel data, long-term meteorological analysis data, and a 3D computer model to produce an FAE analysis that calculates localized peak and return period snow loads on a grid over a specified design geometry.[15] This type of simulation has a long and highly regarded reputation for reliability and is typically used

to aid decisions regarding peak snow load for a building's structural design; however, it can further be useful for the placement and design of snow retention barriers and their attachments.

2. Computational fluid dynamic (CFD) modeling and analysis replicates the wind flows and particle deposition (rain or snow) in the vicinity of or on building surfaces using a 3D computer model in combination with data from a long-term meteorological analyses.[16] This type of simulation is less reliable than an FAE analysis because it is highly dependent on the experience and knowledge of the evaluator. It is therefore typically only used to aid decisions regarding the approximation of relative snow deposition or ice thickness patterns on building surfaces (i.e., light, medium, or heavy zones) rather than specific snow loads or depth values.

3. Computer-aided engineering (CAE) simulations are developed using 2D or 3D computer models to solve the trajectory path, landing zone, and impact force of a sliding ice sheet or snow mass that can release from a sloped roof or facade.[17–19] These types of simulations are typically used to determine the potential for damage and peak impact forces on roof-mounted retention barriers or equipment, building components, and lower roofs, as well as for the determination of landing zones at grade. Similar to the CFD style of analysis above, this type of analysis is highly dependent on the experience and knowledge of the evaluator, and therefore a single output should be treated with caution. A best practice method would be to perform a sensitivity analysis on the various inputs, treating each output as one possible scenario in a range of possibilities that could occur.

- *Field investigations and research studies*: Field investigations are associated with incident or storm event investigations, where the details of problem conditions are investigated and documented, including the gathering of eyewitness observations, dates, current weather data, and so on, used to determine the likely cause, severity, and estimated future frequency of occurrence of similar events. However, field investigations also include outdoor research activities such as the long-term observation of specified mock-ups,[20] the photographing of other local similar buildings that are in the vicinity of the project, or the investigation and documentation of ice and snow collection and performance on existing buildings that are catalogued for future reference.

CASE STUDY: EVALUATION OF COMMON FACADE ARCHITECTURAL ELEMENT

The following case study evaluates a common facade architectural element that is described as a protruding ledge, which historically has been known to produce hazardous falling, sliding, or wind-released snow, ice, and icicles from buildings. This element is considered to be an example of a broader range of facade elements that are susceptible to ice and snow release, such as window sills, horizontal mullion caps, horizontal or vertical fins, solar shading devices,[21] and so on. Photographs of architectural variations of the protruding ledge concept are provided in **figure 1**. **Table 1** provides photographs of the ice and snow problem condition on additional protruding ledge examples, as well as contrasting discussions of three approaches to risk avoidance and mitigation, within the context of the protruding ledge problem condition.

FIG. 1 Examples of window sill, horizontal mullions or ledge details.

TABLE 1 A case-study evaluation of common facade architectural elements

Protruding ledge problem condition:

1. A low-slope ledge with a top surface that projects 12, 18, 24, 36, and even 48 in. or more.
2. The larger scale of this surface allows for drifting and overhanging snow, ice masses, and icicles that can release.
3. A mitigation strategy is required that retains snow and ice, controls overhanging snow, and contains a method of melt-water drainage mitigation to reduce the potential for icicle formation, as well as the potential for the wind-lifted release of ice and snow.

(continued)

TABLE 1 A case-study evaluation of common facade architectural elements (*continued*)

Example photographs of the protruding ledge problem condition:

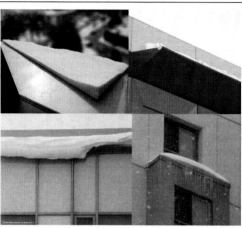

Contrasting Approaches to Risk Avoidance and Mitigation

High Operational Risk Avoidance	Moderate to High Operational Risk Avoidance	Low to Moderate Operational Risk Avoidance
• Typically required by designers and owners when there is a critically sensitive area below or an incident has occurred. • Requires mitigation that provides a high level of confidence that snow, ice, or icicles of any magnitude will not be released except under the most extreme and unusual events.	• The most commonly chosen strategy that allows for some architectural freedom by designers while mitigating the areas with the highest potential for hazard. • Reduces the burden on operations staff to monitor and manage ice and snow on the building. • However, it still must be accepted that temporary restricted access will be required under certain infrequent circumstances.	• Typically chosen when the area below the ledge is not occupied, the ledge is not a high elevation, and there are no skylights, equipment, or property below that could be susceptible to damage. • Mitigation is focused on avoiding widespread releases of ice and snow that could damage a lower roof and create an unacceptable impact noise or the potential for a sliding or windblown falling trajectory beyond the building or property line. • This level of mitigation places a substantial burden on operations staff to monitor and manage the building because temporary restricted access will likely be required more frequently as smaller pieces of snow, ice, and icicles can still release and fall, traveling a substantial distance under the influence of wind.

(*continued*)

TABLE 1 (continued)

Example Mitigation Strategies

High Operational Risk Avoidance	Moderate to High Operational Risk Avoidance	Low to Moderate Operational Risk Avoidance
• Mitigation in the form of a significant application of heat, a dramatic modification of design, or a fully encompassing retention strategy, that is validated to address sliding snow, falling icicles and/or wind released snow and ice under negative wind pressures. • The above strategies are often found to be undesirable from design aesthetic and cost perspectives during design.	• Mitigation in the form of a minor design modification intended to reduce the magnitude of ice and snow, or the application of previously validated continuous snow and ice barriers or fences to retain sliding ice and snow.	• Mitigation in the form of individual snow guards, pins, or ice-breakers that are applied in a distributed pattern to break up ice and snow as it releases. • This mitigation method is typically reserved for situations in which a secondary gutter, lower roof, canopy, or inaccessible area exists below, providing added protection for people and property.

Conclusions and Recommendations

It has been shown that the methods of evaluating architectural designs to identify and mitigate potential falling, sliding, or wind-released snow and ice from buildings is a highly experience-based effort that relies on the past knowledge and expertise of the evaluator. Furthermore, the evaluator requires the skills necessary to: identify potential design impacts, effectively communicate risk to decision-makers, determine the relative effectiveness and feasibility of the available mitigation and management strategies, and most importantly achieve the level of risk avoidance specified by the design team or ownership. This includes the determination of the criteria by which the risk avoidance will be evaluated in the absence of applicable industry guidelines, standards, and building codes.[11]

On the basis of this conclusion, it is proposed that the building industry urgently work to develop guidelines and standards specific to ice and snow accumulation and release from building roofs and facades that will assist and guide the future advancement of building codes, as well as influence the evaluation and mitigation of risk by building designers and owners for current building projects. As such, it is recommended that the design industry actively debate the following fundamental questions:

- What constitutes a hazard with respect to falling ice or snow from a building?
- What criteria or metric should be used when assessing safety?
- What is an acceptable time frame or frequency of temporary restricted access?

ACKNOWLEDGMENTS

I wish to acknowledge the hard work and dedication of the E06.55.13 Task Group--Evaluation of Snow and Ice Accretion on Buildings and Structures as part of the ASTM International Subcommittee E06.55 on Performance of Building Enclosures.

References

1. *Minimum Design Loads and Associated Criteria For Buildings and Other Structures*, ASCE/SEI 7-16 (Reston, VA: American Society of Civil Engineers, 2017).
2. *Bases for Design Of Structures—Determination Of Snow Loads On Roofs*, ISO 4355:2013 (Geneva: International Organization for Standardization, 2013).
3. C. J. Williams, M. Carter, F. Hochstenbach, and T. Lovlin, "Sliding Snow and Ice on Buildings: A Balance of Risk, Cost and Aesthetics," in *Snow Engineering V: Proceedings of the Fifth International Conference on Snow Engineering*, ed. P. Bartelt, E. Adams, M. Christen, R. Sack, and A. Sato (London, UK: Routledge, 2004), 59–64.
4. M. Carter and R. Stangl, "Falling Ice and Snow—Mitigating Your Facility Risk," International Facility Management Association, February 3, 2017, http://web.archive.org/web/20210326184209/https://fmcc.ifma.org/Falling-Ice-Snow-Mitigating-Facility-Risk
5. E. Gurnon, R. Davis, and S. Rhodes, "Ice From Michigan Ave. Building Falls, Kills Man," *Chicago Tribune*, March 1, 1994, https://web.archive.org/web/20220228182253/https://www.chicagotribune.com/
6. N. Norris, D. Andre, P. Adams, M. Carter, and R. Stangl, "The Implication of Energy Efficient Building Envelope Details for Ice and Snow Formation Patterns on Buildings" (paper presentation, 14th Canadian Conference on Building Science And Technology, Toronto, October 28–30), https://web.archive.org/web/20220228182859/http://obec.on.ca/sites/default/uploads/files/members/CCBST-Oct-2014/B1-3-a.pdf
7. T. Janoski, A. Broccoli, S. Kapnick, and N. Johnson, "Effects of Climate Change on Wind-Driven Heavy-Snowfall Events over Eastern North America," *Journal Of Climate* 18 (2018): 9037–9054, https://doi.org/10.1175/JCLI-D-17-0756.1
8. National Weather Service, "Biggest Snowstorms (One Foot or More) at Central Park (1896 to Present), May 9, 2021," https://web.archive.org/web/20220228183224/https://www.weather.gov/media/okx/Climate/CentralPark/BiggestSnowstorms.pdf
9. M. Carter and R. Stangl, "Caution: Falling Ice and Snow—What Is An Acceptable Risk?" (paper presentation, Proceedings of the 7th International Conference on Snow Engineering, Fukui, Japan, June 6–8, 2012).
10. T. Anderson, "Codes and Roof Snow Retention," International Institute of Building Enclosure Consultants—Interface (March 2005), https://web.archive.org/web/20220228184221/https://iibec.org/wp-content/uploads/2016/04/2005-03-anderson.pdf
11. O. Jones, "Falling Ice Capades: Extell Sued Over Icicles at Central Park Tower," *The Real Deal*, December 11, 2020, http://web.archive.org/web/20210326191152/https://therealdeal.com/2020/12/11/Falling-Ice-Capades-Extell-Sued-Over-Icicles-At-Central-Park-Tower
12. Metal Construction Association, "Qualifying Snow Retention Systems for Metal Roofing," 2018, https://web.archive.org/web/20220228184551/https://metalconstruction.org/view/download.php/online-education/education-materials/roofing-educational-materials/metal-roofing-snow-retention
13. *Evaluation Criteria for Standing Seam Metal Roof-Mounted Rail-Type Snow Retention Systems*, EC 029-2018 (Ontario, CA: International Association of Plumbing and Mechanical Officials, 2018).

14. G. Heydari, "Toward Anti-Icing and De-Icing Surfaces: Effects of Surface Topography and Temperature" (PhD diss., KTH Royal Institute of Technology, 2016).
15. A. Brooks, J. Bond, S. Gamble, and J. Dale, "Parametric Simulation of Roof Structural Snow Loads" (paper presentation, Resilient Infrastructure, London, June 1–4, 2016).
16. J. Beyers, X. Qiu, and R. Hakimi, "Meteorological Prediction and CFD Simulation of In-Cloud Icing on Tall Buildings" (paper presentation, Proceedings of the 4th International Conference on Snow Engineering)., Trondheim, Norway, June 19–21, 2000).
17. J. Paine and L. Bruch, "Avalanches of Snow from Roofs of Buildings" (paper presentation, International Snow Science Workshop, Lake Tahoe, CA, October 22–25, 1986).
18. Metal Construction Association, "Metal Roof Design for Cold Climates," 2014, http://web.archive.org/save/https://metalconstruction.org/view/download.php/online-education/education-materials/roofing-educational-materials/metal-roof-design-for-cold-climates
19. D. Taylor, "Sliding Snow on Sloping Roofs," *Canadian Building Digest*, No. CBD-228, November 1, 1983, https://doi.org/10.4224/40000674
20. Design and Build with Metal, "Alpine Snowguards Unveils New Research Facility as Part of Intensive Testing Program," January 9, 2020, http://web.archive.org/web/20210326185217/https://designandbuildwithmetal.com/Industry-News/2020/01/09/Alpine-Snowguards-Unveils-New-Research-Facility-As-Part-Of-Intensive-Testing-Program
21. *Standard Test Method for Static Loading and Impact on Exterior Shading Devices*, AAMA 514-16 (Schaumburg, IL: American Architectural Manufacturers Association, 2016).

Sean M. O'Brien[1] and Amarantha Z. Quintana-Morales[1]

Look Out Below! Addressing Falling Snow and Ice Hazards on Building Facades

Citation

S. M. O'Brien and A. Z. Quintana-Morales, "Look Out Below! Addressing Falling Snow and Ice Hazards on Building Facades," in *Building Science and the Physics of Building Enclosure Performance: 2nd Volume*, ed. D. J. Lemieux and J. Keegan (West Conshohocken, PA: ASTM International, 2022), 298–322. http://doi.org/10.1520/STP163520210006[2]

ABSTRACT

Snow and ice falling from building facades can pose serious safety risks to the public and can place architects, owners, and engineers in a position that is vulnerable to litigation. Design trends such as large mullions, projecting trim, sloped or articulated walls and glazing, and an increased use of shading devices provide greater surface area for ice and snow to collect. Improvements in building enclosure energy performance can also result in colder exterior surfaces on which precipitation can freeze. Potentially as important, increasingly severe and erratic climate patterns (more intense storms, faster temperature swings, etc.) can place new stresses on building facades. Despite these risks, there is little consensus in the industry on the best strategies for mitigating the buildup and shedding of these winter hazards or on standards for testing a facade's performance as it relates to falling snow and ice. This paper evaluates several mechanisms that enable snow and ice to accumulate on and fall from buildings. Using traditional snow retention designs for roofs as an introduction, we focus on how snow and ice interact with facade elements and discuss how contemporary design trends, improvements in performance, and changes in climate patterns contribute to the problem. We review the lack of industry guidance for designing and evaluating snow and ice retention systems. We present several case studies, including research and laboratory testing and project-specific mock-up testing

---

Manuscript received January 8, 2021; accepted for publication October 6, 2021.
[1]Simpson Gumpertz & Heger, 525 7th Ave., 22nd Floor, New York, NY 10018, USA S. M. O. https://orcid.org/0000-0002-4157-871X, A. Z. Q.-M. https://orcid.org/0000-0001-7315-4527
[2]ASTM Second Symposium on *Building Science and the Physics of Building Enclosure Performance* on April 24–25, 2022, and June 12, 2022 in Seattle, WA, USA.

Copyright © 2022 by ASTM International, 100 Barr Harbor Drive, PO Box C700, West Conshohocken, PA 19428-2959.

ASTM International is not responsible, as a body, for the statements and opinions expressed in this paper. ASTM International does not endorse any products represented in this paper.

that our firm has performed. We discuss how the results from this work can help inform the development of design strategies to mitigate snow and ice issues on building facades. The primary goal of this paper is to help build consensus on the need for more industry guidance to address snow and ice hazards on building facades and present some first steps toward the development of this guidance.

Keywords

snow, ice, accumulation, accretion, shedding, building, facades

Introduction

In recent years, cases of ice and snow falling from buildings have received increasing attention for causing significant disruption in busy, largely pedestrian cities. Walking around Chicago or New York in the winter, it seems like everywhere one turns there are caution signs warning to steer clear of falling ice and snow hazards from buildings above. This suggests that the way we are currently designing buildings, changes in climate patterns, or a combination of the two, are contributing to an increasingly prevalent problem.

Falling ice not only creates a risk for pedestrians but also places owners, architects, and engineers in a position that is vulnerable to damage claims and litigation. However, surprisingly little (to no) industry guidance exists for the identification of at-risk elements or the design of snow and ice retention systems for building facades. Designers, peer reviewers, and even owners may intuitively recognize surfaces that could be vulnerable to accumulation but have no standards, design guidelines, or building code references to guide them on how to modify the design to prevent snow and ice from falling or minimize the associated risks.

Building codes and standards provide some guidance when it comes to snow and ice accumulation on roofs, which can serve as a reference for understanding accumulation on facades. Most industry literature focuses on snow and ice accumulation and designing for the resulting dead loads but does not cover what should be done to prevent the snow and ice from falling to adjacent pedestrian walkways and creating a hazard. Through experience, designers and consultants have come to better understand the processes that cause ice dams and icicles to form at roof edges and have developed design strategies to mitigate this specific issue. However, the design of other mitigation strategies, such as surface-applied heat technology or retention devices, is largely unregulated. Manufacturers have led the way by providing some guidance for the design and layout of their proprietary systems; however, these guidelines cannot always be directly applied to facades, especially as the retention devices are intended to be mostly out of view on a roof and rarely fit within architects' design intent for highly visible facades.

A review of industry literature on the topic of falling snow and ice on building facades shows that most research focuses on understanding how and where accumulation will occur on a building and that recommendations focus on eliminating these conditions. Nevertheless, for a variety of reasons, including aesthetics and energy performance, building facades continue to be designed and built with elements that contribute to the risk of falling snow and ice. Where recommendations to mitigate this risk are provided, they fall back on the same limited options typically applied to roofs, including surface-applied heat technology such as heat trace, coatings, or, more generically, retention devices. The design of the retention devices is again left up to the architect, without guidance as to what works and what does not and with limited precedents to review for reference.

It is clear the industry understands this is an issue that requires further attention. For example, recent standards have begun to look at evaluating the loads and impacts of falling ice or snow on exterior horizontal shading devices. ASTM International has initiated a new Task Group E06.55.13, within ASTM Subcommittee E06.55 on the Performance of Building Enclosures, to focus specifically on snow and ice accretion. Nevertheless, much of the current responsibility for protecting the public from the hazard of falling ice and snow is placed on building managers who are encouraged to perform regular inspections to determine the risk of falling ice and snow and to restrict access to certain areas when this risk is ascertained. These strategies focus on identifying and addressing the symptoms rather than the underlying problem.

Falling snow and ice risks are gaining attention and will lead to increased liability for designers in an area that is still not well understood. As consultants, designers, and building professionals, we must strive to better understand the factors that contribute to snow and ice accumulation and shedding on building facades. Understanding these mechanisms, through a review of the literature, laboratory testing, and project-specific mock-up testing, can help define the problem and inform the development of practical design guidelines for how to mitigate snow and ice issues on building facades.

Snow and Ice on Roofs

A review of the mechanisms of snow and ice accumulation on roofs, as well as the limitations of existing industry guidance, can help set the stage for understanding accumulation on facades. The building industry has a long history of studying snow and ice buildup on roofs. The primary snow and ice risks for roofs include increased loads, formation of icicles and ice dams at roof edges, sliding ice and snow, wind-blown ice and snow, and icy walkways, stairs, or ramps adjacent to roof edges (this last one being more a consequence of the previous). These events can lead to personal injury and damage to building elements, landscaping, adjacent construction, sidewalks, cars, and so on. Sloped surfaces, complex geometries, and canopies are particularly vulnerable to sliding snow and ice dams/icicles, whereas low-sloped surfaces are more vulnerable to damage due to snow loads.

ASCE/SEI 7, *Minimum Design Loads for Buildings and Other Structures*, serves as the basis for most building codes when designing for snow and ice loads. Chapter 7 includes procedures for calculating snow loads on flat or low-sloped and sloped roofs on the basis of predetermined site-specific ground snow loads, exposure, slope, building thermal conditions, and risk category.[1] This chapter also looks at increased localized loads that can occur from snowdrifts that shift snow from higher roof surfaces, adjacent structures, or other terrain features, and sliding snow loads caused by snow sliding from an upper roof onto a lower roof. Snowdrifts can create significant loads, often high enough to cause localized failure of the roof structures as shown in **figure 1**.

Chapter 10 of this standard discusses atmospheric ice loads due to freezing rain, snow, and in-cloud icing. It provides guidance for designing "ice-sensitive structures," defined as "structures for which the effect of an atmospheric icing load governs the design of part or all of the structure,"[2] for example, lattice structures, light suspension and cable-stayed bridges, open catwalks and platforms, and signs. In addition to accounting for increased dead load due to ice accretion, this chapter includes guidance for increasing wind loads due to the higher projected area of the structure with the additional ice accretion. It is worth noting that although the effects of accumulated ice are dealt with in detail, the methods of accretion and risk of falling are not (which is understandable given the intent of this standard).

A secondary risk at roofs is ice dam and icicle formation at eaves of sloped roof surfaces. This is an area where current building codes and standards provide little to no specific guidance, although the mechanisms that cause ice dam and icicle formation are relatively well understood by the industry, and designers and consultants have developed mitigation strategies that have largely proven successful. As stated by the U.S. Army Corps of Engineers' *Commentary on Snow Loads*, Technical Instruction (TI) 809-52, ice dams and icicles tend to form at the eaves of "inadequately insulated and ventilated roofs of heated buildings."[3] Sloped roofs that lack insulation at the attic floor level allow heat from the building interior to

FIG. 1 Partial roof collapse due to drifting snow (outlined).

reach the sloped roof surfaces, melting accumulated snow on the roof surfaces. As this meltwater flows to the unheated roof eaves that project beyond the thermal envelope of the building, it refreezes, creating ice dams, and icicles. The more meltwater that collects, the larger the ice dams and icicles become. Eventually this water can back up and leak through the roofing system, if not adequately protected, or create a large enough buildup that the snow, ice, or both break off and fall to the ground below. Although the proper use of waterproof underlayments can prevent water leakage at ice dam locations, large icicles pose a significant hazard to both building occupants and pedestrians, especially when located near building entrances (fig. 2).

Providing insulation and an air barrier at the attic floor level reduces the amount of interior heat that reaches the sloped roof surfaces. Eave and ridge vents allow any warm air that does get into the attic, through breaches in the insulation or air barrier, to escape from the attic. These design modifications result in "cold roofs" that allow the snow to collect on the roofs more evenly, minimizing end dams and icicles at the edges. Guidelines for venting roofs and cathedral ceilings are well established in the industry, with dozens of conference papers and journal articles readily accessible to the average roof designer.[4]

Even with myriad industry guidance, there are two issues that the above strategies do not directly address. The first is the issue of ice and snow (which forms in a manner to prevent icicles) sliding off or being blown off roof surfaces and onto areas below. The cold roof strategy still allows for snow to accumulate on the roof surfaces, and eventually some or all of this snow must melt or come off these surfaces. Sliding snow can damage roof mechanical and plumbing equipment such as vent stacks, damage standing seam roofs, and even rip off gutters or fascias.

FIG. 2 Heavy ice accumulation over a residential building entrance.

Depending on the height of the building, sliding or wind-blown snow and ice can also impose large dynamic loads on objects below.

The second issue is that of differential solar heating of roof areas. Cold roof design is highly effective on large sloped roofs with little to no shading or detail. However, for roofs with dormers and multiple pitches, two problems arise. First, with complex roof geometries, it can become difficult or impractical to maintain a continuous ventilation path from eave to ridge, resulting in localized unvented areas that can lead to melting snow. Second, ventilation design is typically based on preventing melting snow caused by heat losses from the building interior. Even when exterior temperatures are well below freezing, solar gain on snow-covered roofs can lead to very high temperatures and result in snowmelt regardless of venting. As noted above, on a clear pitch this may not be problematic, but on a complex roof on which multiple elements below the melting area are shaded (i.e., below freezing), ice formation down slope can be a significant issue. This is exacerbated by the intensity of the melting, which for an area with high solar gain and minimal snow cover can be orders of magnitude greater than what would occur due to building heat losses alone. The building entrance shown in **figure 2** was compromised for this very reason—the roofs of the building were all vented, but dormers and other roof features created differential shading that led to the formation of significant amounts of ice at the roof edges and on surfaces below.

The industry has turned to surface-applied heat technology and retention devices to reduce the effects of sliding and wind-blown snow and ice by either melting the accretion or retaining it until it can melt or shed in smaller pieces. However, this is again an area in which codes and standards are limited. Surface-applied heat technology typically includes heat trace cables installed along the roof (most commonly at the edge) to melt accumulated snow and ice and create paths for the meltwater to drain. The latter is a critical aspect of the system because in freezing conditions meltwater needs a complete path to an outlet to avoid refreezing and meltwater backups, which can create additional hazards or overload components. The amount of cables and their layout on the roof surfaces is usually determined on a project-by-project basis by the designer and manufacturer. Many times, these systems are installed after the fact by building managers or owners—as a remedial approach once a problem has been identified as opposed to a proactive design measure. The same can occur with retention devices. TI 809-52 recommends providing snow guards (i.e., retention devices) on slate and metal roofs, barrel vaults, and other roofs with smooth membranes.[3] It recommends multiple rows of snow guards, spaced "well apart."[3] For calculating loads, it recommends assuming that the friction between the snow and roof is zero, designing guards to resist the snow located within 45° up slope, and using a design load that is less than half of the manufacturer's reported failure load, at a maximum. In the end, TI 809-52 acknowledges that "design guidance, test data, and performance standards on snow guards are limited so they should be used with caution."[3] We found that much of the guidance for retention devices comes from the relatively few manufacturers that supply them. Several manufacturers[5,6] have helpful publications that discuss the historical use of snow guards and why and how to use them. Some also offer design assistance for their proprietary systems. Their recommendations are generally in line with those of TI 809-52.

Heat trace and retention devices are not perfect solutions, and a lack of standardization can make it difficult predict or evaluate the performance of each system. Heat trace requires coordination with the building's electrical system and, as mentioned above, forethought as to how to drain the meltwater. Systems must be adequately anchored so that sliding snow does not rip them out before they are able to work. Retention devices must also be engineered, and many times tested, to verify attachment, and the attachment methods designed to be appropriate to the type of roof onto which they are installed to avoid damaging the roof. Some systems rely on adhesives, which may not be reliable long term in some climates. Ultimately, codes and standards are needed to better guide designers and consultants on how to design mitigation strategies for sliding and wind-blown snow and ice. The industry has taken notice of this deficiency. Subcommittee Task Group E06.55.13 is working on developing a standard for the design of snow retention devices on roofs. As it stands, with no code-mandated requirement to evaluate and address these issues on all projects, building owners are often left having to manage this risk on their own, restricting access to areas below potential falling hazards or removing the snow and ice themselves.

Snow and Ice on Facades

The issue of snow and ice falling from facades is by no means a recent development. In reviewing news sources and historical references, we found newspaper articles dating as far back as 1903 citing damage (and even death) resulting from ice falling from buildings. In one instance (New York City, 1903), there is a direct reference to the lack of a "city ordinance which provides for the removal of such ice to insure [sic] the safety of pedestrians."[7] The authors began investigating snow and ice issues on facades around 2014 in response to both project needs and an apparently increasing frequency of falling snow/ice events in cities in which our company has offices. We say "apparently" because it is not possible to determine whether this increase is due to an actual increase in events, increased media coverage of events (including less "mainstream" outlets that may be more apt to report on these issues, however small they may be), or a combination of the two. Regardless of the cause, this increased attention has led to many in the building industry questioning whether the designs of their projects will lead to similar issues and asking design professionals to address them proactively. This attention creates potential liabilities for those designers given the relative lack of industry guidance on how to achieve successful snow/ice mitigation designs.

Several design trends potentially contribute to the increased risk of falling ice and snow incidents. Large mullions and projecting trim, sloped walls and glazing, and an increased use of solar shading devices provide greater surface areas for ice and snow to collect. At the same time, an overall increase in building enclosure energy performance, that is, increased insulation, reduced air leakage, and high-performing glazing, results in colder exterior surfaces on which precipitation can collect. Precipitation that accumulates on these surfaces tends to melt where it touches warmer surfaces such as glazing and refreeze as it moves to colder surfaces such as metal framing and, in particular, metal that projects away from the building. This process results in the formation of icicles or ice sheets that can be wind-blown, slide, or fall from the building. Increasingly severe and erratic climate patterns exacerbate the issue by placing new stressors on building facades.

The complexity of the liability issue became clear to us during a project on which we were asked to determine fault for a falling ice issue. In cases of more typical design elements, the relevant building code forms the basic standard of care for the element under consideration. Take, for example, a case in which an architect designs a roof that does not properly slope toward the drains and that roof collapses under heavy ponding water conditions. Regardless of any project-specific mitigating factors, an attorney can simply point to the fact that the building code requires a positive drainage slope on a roof and that the design did not include such a slope. For our project, ice fell from a small (approximately 10 cm) projection on a curtain wall system on multiple occasions, creating a risk for the sidewalks below. The building code, relevant curtain wall design guides, and industry journals/literature provided zero guidance on the issue of falling ice from this type of facade element, and neither the plaintiffs nor defendants in the case could point to a single example of a similar detail in the vicinity of the subject building having the same issue. The issue eventually became overshadowed by larger problems at the building (that were governed by more defined design practices), but the case made clear to us the potential for these problems to create very difficult risk management issues for design professionals if better guidance did not become available.

LITERATURE REVIEW

Ice Accretion on Structures

Studies on the accumulation or accretion of atmospheric ice on structures have historically focused on specialized structures in areas with severe icing conditions such as oil and gas facilities, electrical and telecommunications equipment, ships, and airplanes. These studies provide a basis for understanding how ice and snow accumulation occurs on building facades. The U.S. Army Corps of Engineers' Cold Regions Research and Engineering Laboratory (CRREL) Report 80-31 – Icing on Structures[8] and National Research Council of Canada (NRC) Technical Paper 1968-05 – Atmospheric Icing of Structures[9] categorize atmospheric ice accretion as hoarfrost, rime, or glaze. Hoarfrost forms from the deposit of water vapor on surfaces at or below 0°C and results in minimal accumulation compared with other types of accretion. Rime forms when supercooled water droplets contact a surface and rapidly freeze as "the latent heat of fusion is dissipated."[8] Rime typically forms at temperatures of −3°C to −25°C and can grow as more droplets accumulate on top of the base layer deposited on the surface. In conditions closer to freezing (0°C to −3°C), supercooled water droplets that flow over a surface to create a continuous film before freezing form glaze ice. The type of ice accretion formed is therefore affected by the temperature and size of the water droplets, the rate at which they make contact with the base surface, which is influenced by wind speed and amount of droplets in the air (liquid water content, or LWC), and the air temperature, which affects the rate of heat dissipation and the temperature of the base surface.[8] In general, higher wind speeds at air temperatures close to or just below freezing (resulting in slower heat dissipation) and larger droplet sizes result in mostly glaze ice, whereas lower wind speeds at colder temperatures (rapid heat dissipation) and smaller droplets result in rime. Glaze is harder, has a higher density than rime, and forms a stronger bond with the contact surface.

CREEL Report 80-31 also identifies certain factors that affect the amount of ice that is deposited on a surface. These factors include the surface shape and cross-sectional area, wind speed, air temperature, droplet size, and LWC.[8] Cylinders with a smaller diameter tend to accumulate more ice than those with a larger diameter in part because droplets are more likely to be deflected around the larger cylinder. Lower wind speeds result in increased accumulation because the heat of fusion has sufficient time to dissipate to the surroundings. At higher wind speeds, some droplets do not have sufficient time to freeze before they are carried away by the wind. Lower temperatures result in greater overall accumulation.

CREEL Report 80-31 and NRC Technical Paper 1968-05 list some methods to help control ice accretion while acknowledging that more effective methods are needed. These methods include vibration, heated surfaces, icephobic surfaces, deformable surfaces, and freezing point depressants.[8] Active control methods, the most common and sometimes the only available tools, include mechanically removing the ice with "baseball bats, sledge hammers, axes, hammers, picks, and other impact instruments"[8] or melting the ice using chemical or thermal methods. NRC Technical Paper 1968-05 includes relative values for ice adhesion metals, 88–120 psi, rubber, 20–50 psi, and plastics, 10–40 psi.[9]

Building on the basic principles of ice accretion, a more recent publication titled "Strategies to Prevent Ice Accumulation on Metal Surfaces" summarizes the latest developments in the research into icephobic surfaces. The article acknowledges that researchers are still working to understand the "fundamentals of water-surface interactions," including the "onset of ice nucleation" and "detailed nature of water layers immediately adjacent to a cold surface."[10] The researchers cited in the article, including Ali Dhinojwala and Azar Alizadeh, are studying how water droplets impact and interact with different types of surfaces. Their studies found that although surface temperature greatly affects the rate of water droplet cooling and ice formation, the size of the contact area between the water droplets and the surface also strongly influences the onset and rate of ice nucleation. The researchers are using these findings to evaluate the role that surface structure (flat vs. textured or roughened) and surface chemistry (hydrophobicity) can play in developing anti-icing surfaces. Similarly, Joanna Aizenberg, another researcher in this field, is working with others to evaluate the use of slippery liquid-infused porous surfaces (SLIPS) to mitigate ice accretion. The researchers create SLIPS by "impregnating a porous nanostructured material with a water- and oil-repellent lubricating liquid," a process inspired by the pitcher plant, whose "microscopically rough surface … locks in a smooth lubricating coating of water."[10] Aizenberg was part of a recent study that evaluated a SLIPS-treated aluminum and found that it outperformed untreated aluminum during high humidity, freezing, and melting tests. The treated aluminum remained ice free during most of the test, and any ice that did accumulate melted quickly, whereas the bare aluminum accumulated more ice, which stayed in place for a longer period of time. The applicability of superhydrophobicity in the field of building integrated photovoltaic (BIPV) design has also been reviewed.[11] Additional research and development is needed to verify the commercial application of these ice mitigation strategies, but they appear to be promising alternatives.

Until recently, climatological data as relates to ice accretion have been limited. This makes it difficult to evaluate or predict incidences or amount of ice accretion on the basis of geography, weather, or seasons. As described in "Quantitative Ice Accretion Information from the Automated Surface Observing System," in the early 2000s the National Weather Service approved the implementation of a new algorithm that would allow the Automated Surface Observing System (ASOS), the primary surface weather observing system in the United States, to report quantitative estimates of ice accretion.[12] The intent of this project was to use the ASOS system to provide a source of more standardized and reliable ice accretion data than had thus far been available. As more of these data are aggregated and evaluated, designers will be able to better understand the conditions to which they should be designing, and owners and building managers will be able to better prepare for ice events.

Ice and Snow on Facades

Looking more specifically at the topic of falling ice and snow on facades, there are a few recent publications that are worth noting. AAMA 514-16[13] provides a standard laboratory procedure for evaluating the loads and impacts of falling ice or snow on exterior horizontal shading devices. The test method uses sandbags to simulate static loads from ice and snow accumulating on the shading device and a cylindrical ice compactor to simulate impact loads falling onto the shading device. This standard provides valuable guidance for understanding these loading conditions while acknowledging that snow and ice accretion should be considered in the design of these systems. The test method specifically states, however, that it does not evaluate the effectiveness of the shading device to prevent release of the original ice or snow on the shading device. Further, Note 10 states, "AAMA recommends that the design or facility management team limit pedestrian and/or vehicular access to otherwise unprotected areas below exterior shading devices whenever the potential for snow or ice buildup exists,"[13] ultimately placing responsibility for addressing the falling ice and snow back on the designer and building owner.

CRREL published a critical report that provides a first step toward understanding the effectiveness of retention devices on exterior wall fins. The report documents testing performed for a new sports stadium in the Northeastern United States.[14] Recognizing the potential hazard of ice and snow sliding from exterior wall fins installed at the stadium perimeter to the expectedly large crowds attending football games, concerts, and other entertainment events, the testing objective was twofold. The first goal was to determine whether this hazard existed on the as-designed horizontal fins. The second was to evaluate the performance of four types of ice retention devices (IRDs) in mitigating this risk. Three of the IRDs were continuous strips with varied profiles, and the fourth included intermittent truncated cones spaced at 30 cm on center. All four IRDs were located at the fin edge and spaced off the fin using 0.5-cm spacers to provide for drainage. Testing of the unmitigated fins showed that, with the fins located at both 30° and 50° from the horizontal, the accumulated snow and ice slid off the fins in a single large piece (approximately two thirds the size of the 4-m fin), creating a significant safety risk. All the IRDs were effective in retaining the snow and ice. However, at each of the continuous IRDs, the drainage path below the strips became obstructed as the meltwater froze, causing additional meltwater to drain toward the ends or eventually over the retention devices and form large icicles. The intermittent IRDs, or buttons, were not only effective in retaining the accumulation but also allowed the meltwater to drain in between without leading to icicle formation. The buttons provided a mechanical locking of the accumulation by allowing the snow and ice to form around the shape. Ultimately, snow and ice did slide off between the buttons, but the size of these pieces was limited by the button spacing. Icicles that did form on the front edge of the fins were also relatively smaller in size to those created at the fin ends with the continuous strips. Overall, the testing demonstrated that the probability of a hazard was relatively high, given the size of the ice sheets that fell from the fins, and that IRDs should be installed to mitigate this hazard. The intermittent IRDs were the most effective because they retained the accumulation without creating the additional hazard of large icicles at the edges. This study, which is publicly available, is an excellent resource for designers facing similar architectural features on their buildings.

There are few other publications or resources in the industry literature that focus specifically on this topic. Some specific articles have been written regarding snow and ice accumulation on tall buildings[15] but, as we have found, tall buildings are not the only potential source of hazards in urban environments. The literature in this area is still in its infancy, requiring significant input from the design and construction community in order to mature.

CHALLENGES

The advantage of most roof snow and ice mitigation approaches is that the elements in question are typically not visible, or are minimally visible but not aesthetically objectionable or tend to blend into the surrounding roof areas. Snow and ice are also most often a problem on steep-slope roofs. The majority of these types of buildings occur in residential areas or, if in urban areas, are set back from street fronts and sidewalks. Buildings in urban areas typically have low-slope ("flat") roofs that are not subject to ice damming or sliding snow, and any steep-slope roofs are usually set back from the edge of the building. This makes falling snow and ice from roofs much less likely to cause personal or property damage and in many cases limits the need for protection to areas near building entrances/exits.

The issue of snow and ice falling from facades introduces three significant complications. First, the surfaces at risk of snow and ice accumulation typically abut sidewalks, streets, or other public thoroughfares. In urban environments, at-risk surfaces may be at significant heights, which means even small falling objects can carry significant energy. For example, in the winter of 2014, an approximately 1-kg chunk of ice separated from a building facade in New York City and fell over 300 m before hitting a sidewalk, right outside a public transit hub. The ice struck the sidewalk at over 45 m/s and resulted in the shutdown of the hub, affecting the morning commute of tens of thousands of people. Fortunately, no one was struck by the falling ice; to cite a more recognizable example, being struck by that falling piece of ice would result in twice the impact force of being hit directly on the head by a major league baseball pitch. While moderate heights pose a risk to those directly below, they are also problematic because falling snow and ice can "drift" with the wind and affect sidewalks or streets far from the source.

The second challenge lies in mitigation approaches—more specifically, the kind of retention devices or heat trace cables often used in roof applications. These elements are very difficult to seamlessly integrate with facade systems in a way that does not impact aesthetics, both from the exterior and (for occupants) the interior. Further, active systems such as heat trace or more unique options such as hydrophobic coatings or "gripper" surfacings are not reliable long term and will require maintenance or reapplication over time. Maintenance and repair become a major issue for a 50-story building, or for an active heating system that is integral to facade elements that are not easily removed. Because of these limitations, mechanical retention devices and designing to avoid susceptible surfaces are much more reliable approaches to minimizing risks.

The third complication when dealing with snow and ice on facades is that there are multiple mechanisms by which ice can form on buildings. While melting and refreezing of snow is probably the most common mechanism, building facades may also be subject to ice formation due to freezing rain exposure. Freezing rain can cause ice to form on almost any surface, vertical or horizontal, as opposed to snowfall, which will only impact near-horizontal elements. Atmospheric icing, forming rime, or glaze ice as described above can affect super-tall (>300-m) buildings in predominantly cold climates. This is a relatively new issue for buildings in the United States as, until recently, the tallest buildings in the world were built in much warmer climates such as the Middle East and Southeast Asia.

THERMAL MODELING

We performed thermal modeling of several building enclosure details to evaluate the potential for the melting/refreezing of snow, leading to the development of ice formation. This was part of our initial effort to better understand the mechanisms of ice formation (because understanding the basis of the problem will eventually lead to better solutions—avoidance being one possibility). We used the THERM 7 and WINDOW 7 computer programs form the Lawrence Berkeley National Laboratory to evaluate the steady-state, two-dimensional heat flow through a series of window/wall assemblies. Our initial focus was on the difference between existing, poorly insulated buildings and newer buildings constructed to more stringent energy standards. The motivation for this approach was, as noted above, the apparent increase in falling snow/ice events in the past 10–15 years, suggesting that newer construction may be a contributing factor.

Looking to roof applications for guidance, we posited that one source of ice formation, specifically on projecting elements such as trim, fins, and window sills, was melting of snow on upper surfaces that resulted in meltwater flowing down until refreezing on a cold surface. For a typical building application, the most likely location for this to occur would be below windows. Even thermally broken windows with insulating glass can still remain above freezing on the exterior surface at temperatures when snow is likely to occur (for the purpose of our analysis, we assumed that range to be approximately $-10°C$ to $-4°C$).

We modeled an "old" and a "new" window condition to evaluate the differences between the two. The old model assumed a solid brick masonry wall with limestone sill and a thermally broken aluminum framed window (with insulating glass). The new model utilized insulation meeting the 2015 International Energy Conservation Code (IECC) with the same window as the old. Models were run using an interior temperature of $21°C$ and an exterior temperature of $-4°C$. The basic thermal profiles are shown in **figure 3**. Darker colors/shading (towards the left/exterior side of the figures) represent colder temperatures; lighter colors/shading represent warmer temperatures.

Qualitatively, the old assembly is much warmer on the exterior side due to the lack of insulation in the stone sill, which allows heat to bypass the thermal breaks in the window system. We then evaluated surface temperatures on the exterior of the window glazing and framing, as well as the sill, to determine where melting and freezing could occur. This is shown in **figure 4**; the white line represents the $0°C$ isotherm.

Figure 4 illustrates the concept that we had suspected, which is that an older, poorly insulated wall loses enough heat to keep exterior temperatures above freezing, even during conditions when snow is likely to form. This is essentially the opposite of the "cool roof" approach, in which above-freezing surfaces will melt and drain snow rather than retain it. For the newer, better insulated wall, the window sill and majority of the wall below remain below freezing due to improved insulation, but a typical insulating glass unit can still allow for melting of incident

FIG. 3 Basic thermal profiles of uninsulated (*A*) versus insulated (*B*) wall assemblies.

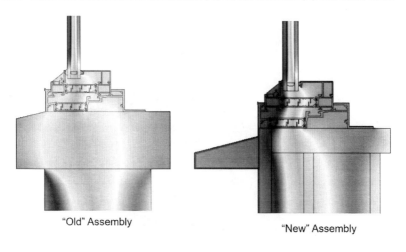

"Old" Assembly   "New" Assembly

FIG. 4 Zones of melting and freezing on window/wall assemblies.

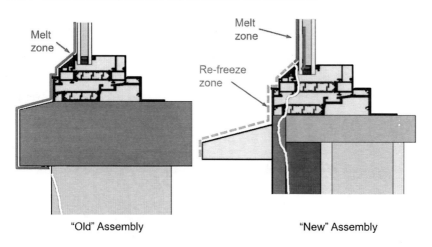

"Old" Assembly   "New" Assembly

snow. Just like an ice dam on a roof, meltwater can run down and freeze/accumulate on the surfaces below.

Although this specific case illustrates the increased susceptibility of better insulated buildings to ice accumulation, it does not cover all cases. We observed an incident of falling ice from the 4th-floor cornice of a building in New York City shortly after completing some initial research on this topic (**fig. 5**).

Several large sheets of ice fell from the cornice of the building as one of the authors was exiting and shattered on the sidewalk, prompting building staff to quickly close off the sidewalk and divert pedestrians. We then performed a study of

FIG. 5 Ice and snow on a projecting 4th-floor cornice.

the impact of snow on the thermal performance of both assemblies discussed above. A loose blanket of freshly fallen snow has a surprisingly high R-value (13 cm of snow gives an approximate RSI value of 0.5). This is due to the cellular structure of the loose snowpack, containing thousands of tiny air voids that result in reduced heat flow. The addition of this "insulation" shifts the thermal balance of the assembly. Whereas the previous iteration of the new wall showed consistent freezing conditions on the near-horizonal surfaces, the addition of snow results in above-freezing temperatures below the snow (**fig. 6**).

FIG. 6 Impact of snow accumulation on "new" window sill.

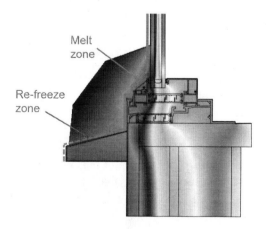

The introduction of snow creates a condition in which snow accumulation, which cannot be practically prevented on near-horizontal surfaces, will result in ice formation over time. This result indicates that for projecting fins near windows, it may not be possible to prevent snow and ice accumulation, making retention the only viable strategy.

Following this analysis, we performed a series of laboratory tests at CRREL in Hanover, New Hampshire, as described in the following section, to further evaluate our analytical findings in a more realistic environment.

LABORATORY TESTING

We designed a laboratory experiment to further study the findings from our analysis. The mock-up assembly, intended to be tested in a climatic test chamber at CRREL, consisted of a code-compliant stud-framed wall system, thermally broken aluminum-framed strip windows, and a series of projecting aluminum fins of various depths and distances from the window sills. We also included an uninsulated precast concrete window sill on one end of the mock-up to better evaluate the old assembly variation that we modeled. The overall mock-up is depicted in **figure 7**.

We subjected the mock-up to multiple rounds of snow exposure, using artificial snow "guns" similar to those used at ski resorts. This is shown in **figure 8**. The goal was to maintain a constant "interior" temperature of approximately 21°C and "exterior" (cold room) temperature of −4°C to replicate the conditions used in our thermal modeling. However, we had to make some adjustments to the cold room temperature for snow generation to occur. For the first few days of testing, the cold room temperatures cycled from approximately −10°C to −30°C at the start of testing to just below 0°C toward the middle of the day and back down at the end of the day to prepare the room for snow generation for the next day. After some trial and error, the mock-up temperatures stabilized at approximately 24°C interior and

FIG. 7 Mock-up assembly. The "old" window sill is on the left side.

FIG. 8 Mock-up assembly after being subjected to artificial snow exposure.

−1°C exterior. We monitored the performance of the mock-up through the multiple temperature cycles, as well as during artificial "sun exposure" to approximate winter solar exposure. The "artificial sun" was composed of a 5-ft.-diameter array of thirty-one 150-W light bulbs for a total of 4,650 W and was placed 11 ft. and 4 ft. from the mock-up wall.

We were able to confirm many of the theories that we had prior to and during our theoretical analysis. In one case, prior to solar exposure, we directly observed the melting of ice against the surface of the window mullions. We measured the temperature of the mullion immediately after shedding of a 2-in.-thick section of ice to be 0.5°C, whereas a similar but fully exposed mullion measured −0.5°C, below the freezing point. As shown in **figure 9**, this confirms our analytical finding that snow/ice accumulating on a surface will change the thermal profile and can lead to concealed melting.

Infrared (thermal) imaging of the mock-up during snow exposure clearly and consistently showed higher temperatures around the old/uninsulated window sill that we simulated (**fig. 10**). Lighter colors indicate higher temperatures in the image.

Although the uninsulated sill was warmer than the surroundings, some ice still accumulated depending on the consistency of the snow. As noted previously, phenomena such as wet snow or freezing rain can still produce ice on surfaces that are at or just above the freezing point. We did find that under melting conditions, snow on the rough-surfaced concrete sill was retained for significantly longer than the adjacent (smooth) aluminum fins. This shows that the surface texture of projecting elements is important to the retention of snow/ice. When ice did shed from the concrete sill, it did so slowly and in small pieces, as opposed to the aluminum fins, where larger sections of ice and snow fell together.

FIG. 9 Melting conditions below accumulated ice on window frame.

FIG. 10 Infrared image of mock-up showing higher temperatures at uninsulated window sill.

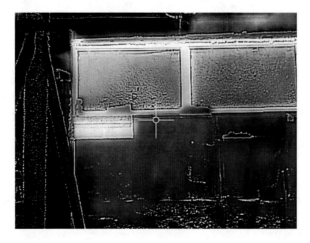

Our mock-up included fins at multiple distances from the windows. The fins further from the windows consistently had lower surface temperatures than those just below the windows, as demonstrated by surface temperature probe measurements taken once the mock-up interior and exterior temperatures stabilized. Surface temperatures of all the aluminum fins were lower than the opaque walls and window glazing, whereas the concrete sill surface was consistently the warmest surface. We observed that accumulated snow and ice tended to slide off the fins closest to the windows, regardless of fin depth. This is consistent with our analytical findings that showed that heat loss at or near windows would impact melting on adjacent components. The fins further from the windows were the last elements to shed snow and ice.

Since our original laboratory analysis, we have had the opportunity to perform testing of actual designed, project-specific building elements in similar climatic test chambers. The design of these experiments needs to consider many factors in order to test as many configurations and orientations of building components as possible. In one instance, this resulted in a mock-up assembly combining multiple aspects of the building details into one compact unit. This unit, which was also being evaluated for water shedding, was mounted on a variable-pitch frame, which was then installed on a turntable within the climatic test chamber. This allowed for both angling and rotation of the assembly and the testing of a large number of options in a fairly efficient manner (**fig. 11**).

The increasing availability of the types of testing facilities that we utilized for this mock-up is a significant benefit to the industry. The ability to perform a wide range of tests on multiple options, and to do so efficiently, will allow more projects to perform this type of testing and improve the body of knowledge about falling snow and ice in the building industry.

FIELD MOCK-UP TESTING

The sections above focus primarily on addressing snow and ice hazards in new construction, but as noted previously, existing buildings make up the vast majority of the building stock in most areas. Our firm has been involved in some projects on which snow and ice hazards developed on existing buildings and required remediation. As with many aspects of the building industry, such as building code development, there is often a disproportionate focus on new construction. The evaluation of snow and ice problems in existing buildings can present unique learning opportunities for the industry, both in terms of identifying features that lead to problems and evaluating solutions. Many features of building design, such as blast-resistant fenestration, dry floodproofing of building areas, and even facade components designed for extreme wind events, will never be fully tested due to the infrequency of such extreme events. Snow and ice issues fall into a similar category—the practical reality is that many of these systems will never see exposure to their design conditions due to the difficulty of predicting where and when those conditions occur. An existing building with known problems provides a unique opportunity to test mitigation strategies in an area in which design conditions have been confirmed.

FIG. 11 Laboratory testing of building enclosure mock-up.

Our firm provided consulting services for a building in the Northeastern United States where snow and ice were accumulating and falling from architectural features (fins and ledges). We developed, partly on the basis of the research and background information presented above, a series of options aimed at retaining snow and ice and allowing it to slowly melt and dissipate. Options included discrete buttons, continuous (slotted for drainage) rails, and surface applied "gripper" strips to improve adhesion of ice to the sloped surface (**fig. 12**).

Findings from this evaluation, which included analysis of time-lapse footage during and after snowfall events, demonstrated that the discrete snow buttons provided the best performance, leading to their widespread implementation throughout the building.

PRACTICAL APPLICATIONS

During the course of our research we were involved in a building project for which the issue of potential ice hazards was raised, both by our team and by the owner. This building included a large number of sloped surfaces at window sills, similar to

FIG. 12 Field mock-up of snow and ice retention devices.

surfaces where we directly observed falling ice on a separate (existing) building. We assisted the building architect in the design of retention devices for these sloped surfaces. We based the design on the research from CRREL as well as our thermal analysis and mock-up testing, which identified projecting elements below windows as the most susceptible surfaces (for this specific application).

The building in question was a 35-story building in New York City. We understood that the geometry of the windows and projections was susceptible to accumulating ice and snow and that the geometry could not be modified due to architectural constraints. In this case, we designed a series of "snow buttons" to be installed at ledges directly below windows. A shop drawing excerpt and image of the actual installed buttons are shown in **figure 13**.

The spacing and setback of the buttons was coordinated with both the architect and owner to reduce the aesthetic impact from both the interior and exterior of the building. The building was completed in 2017 and has so far weathered multiple snow and ice events without incident. This is a good example of the practical implementation of research and past experience that produced a favorable result, adding to the body of knowledge in the industry that will benefit other designers down the road.

CLIMATE AND CLIMATE CHANGE IMPACTS

We reviewed environmental conditions/historical weather data corresponding to multiple falling ice events in the past few years. Many climatic factors can affect the occurrence of falling snow and ice events, including exterior temperatures, cloud cover, solar gain, precipitation, and wind speed. Site geography and exposure, building geometry, and height also impact the effect of these climatic factors on the

FIG. 13 Excerpt from project shop drawings showing snow button (*A*) and photograph of installed buttons (*B*).

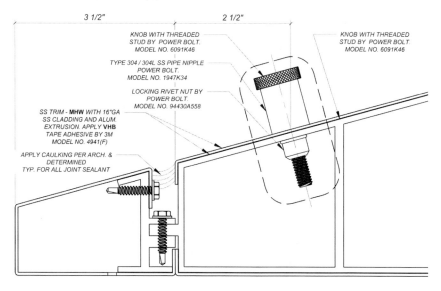

(*A*)

(*B*)

buildings themselves. Some preliminary patterns we have observed in our analysis include a correlation between falling ice and snow events and larger increases in temperature when below freezing (greater than 5°C) or moderate increases in temperature when near freezing (2–3°C). Wind gusts (which are dependent on cityscape geometry and may be intensified in dense urban cores due an increase in friction from buildings) are correlated closely with incidences of ice fall events, more so than an assessment of wind speed alone.

There has been much discussion in the industry of the effects of climate change on buildings, typically related to energy use. However, when evaluating building energy use, it is clear that the types of average temperature increases likely to occur will have little to no impact on energy use. Of greater importance, based on our analysis of weather data and falling ice incidents, is the increase in erratic climate patterns. Snowfall followed by cold or slowly increasing temperatures is unlikely to result in major events, as snow and ice are allowed to slowly melt/dissipate over time. An extreme winter storm followed by days of warm weather can create the opposite, where rapid melting can result in the sudden release of large pieces of ice capable of causing damage or injury. Alternatively, a week of warm weather followed by a sudden snowstorm could result in significant ice formation as residually warm building surfaces lead to melting and refreezing of snow.

Climatic effects on falling ice and snow are potentially significant because having a better understanding of what conditions result in the worst problems can lead to better predictions of where problems may occur. It is important to remember that new construction, where the latest methods for snow and ice mitigation can be implemented, represents a very small fraction of the total building stock in most large urban areas. If municipalities can better predict the conditions that will result in falling snow and ice from unmitigated/existing buildings, building owners will be better equipped to protect the public when risks are the highest.

Conclusions and Next Steps

Our research and practical experience evaluating falling ice and snow from building facades has shown that this is an area with significant need for more research, industry involvement, and standards and building code development. Although we have worked on several projects, both remedial and new construction, on which discrete snow buttons have proven effective, we have also seen many other cases in which more drastic measures (known or unknown) would be necessary. We are still learning about how snow and ice form on buildings, but, unlike areas such as wind and earthquake loading, there are no calculations or procedures to date for quantitatively assessing this risk. This is truly a field that will require the combined observations, efforts, and cooperation of professionals and building owners in all areas of the industry in order to make progress and develop guidelines that will provide real benefit.

Based on our experience, the following are appropriate next steps aimed at improving our collective knowledge regarding falling ice and snow from building facades:

- Increased research into the climatic conditions that most closely correspond to falling ice and snow events. This can aid in the development of predictive models that identify times when falling hazards are the highest. Although not truly preventative, as with many aspects of the industry there is significant benefit to having "advanced warning" systems for hazardous events.
- Further study of specific falling snow and ice events. This will require contributions from a wide range of industry professionals and building designers in order to develop a better understanding of what types of building elements/components pose the highest risk. As discussed in this paper, prediction of falling hazards via calculation or other analytical procedures is extremely difficult. While we work to develop these methods further, practical experience is our best option for furthering our collective understanding of the issues.
- Improved laboratory facilities for testing and evaluation. The facilities that we have worked with have high-quality equipment and trained personnel capable of evaluating the complex issues associated with snow and ice problems on building facades. However, there are very few of these types of facilities, and costs may still be prohibitive for smaller projects—or for projects where a large number of conditions must be evaluated. A good comparison to this situation is that of testing laboratories for window certification. Thirty years ago, when standards for windows and doors were still developing, there were a very small number of accredited laboratories that could perform the relevant testing, which increased both costs and lead times. Today, there are hundreds if not thousands of facilities that can run these types of tests, making them much more accessible and cost-effective. If that were the case for snow and ice testing, the speed of industry advancement could be greatly increased.

ACKNOWLEDGMENTS

We would like to acknowledge the exceptional cold-weather laboratory facilities at the U.S. Army Cold Regions Research and Engineering Lab (CRREL) in Hanover, New Hampshire, and Ontario Tech University in Oshawa, Ontario, where some of our research and project testing was performed, as well as the high quality of the staff at both facilities. We also acknowledge the work of others cited in this paper, as well as of our colleagues in the industry, who have laid the foundation for our ongoing research into the topic of snow and ice on building facades. Thanks also to Trumbull-Nelson Construction Company for assistance with some of our mock-up constructions, as well as Peerless Architectural Windows and Doors for their generous donation of sample windows to support our testing.

References

1. *Minimum Design Loads for Buildings and Other Structures*, ASCE/SEI 7-10 (2010) (Reston, VA: American Society of Civil Engineers, 2010).
2. *Minimum Design Loads for Buildings and Other Structures*, ASCE/SEI 7-10 (2010) (Reston, VA: American Society of Civil Engineers, 2010).
3. *Commentary on Snow Loads*, TI 809-52 (Washington, DC: U.S. Army Corps of Engineers, 1998).
4. W. Tobiasson, T. Tantillo, and J. Buska, "Ventilating Cathedral Ceilings to Prevent Problematic Icings at Their Eaves" (paper presentation, Proceedings of the North American Conference on Roofing Technology, Toronto, ON, Canada, September 16–17, 1999).
5. R. Haddock, "Use of Snow Retention Devices, Science or Science Fiction" (PowerPoint presentation, Ron Blank & Associates, AIA Continuing Education Program sponsored by S-5!/Metal Roof Innovations, Colorado Springs, CO, March 18, 2021).
6. B. C. Stearns, "Designing a Snow Retention System," *Interface* (October 2011): 24–33.
7. "Falling Ice a Menace to Safety of Pedestrians," *Stark County Democrat*, Canton, OH, February 27, 1903.
8. L. D. Minsk, *Icing on Structures*, Cold Regions Research and Engineering Laboratory Report 80-31 (Hanover, NH: U.S. Army Corps of Engineers, 1980).
9. D. W. Boyd and G. P. Williams, *Atmospheric Icing of Structures*, Technical Paper 1968-05 (National Research Council of Canada Division of Building Research, Ottawa, 1968), https://doi.org/10.4224/20378701
10. M. Jacoby, "Strategies to Prevent Ice Accumulation on Metal Surfaces: Fundamental Studies and Coating Strategies Help Avoid Detrimental Buildup," *Chemical and Engineering News* 90, no. 30 (2012). https://cen.acs.org/articles/90/i30/Strategies-Prevent-Ice-Accumulation-Metal.html
11. P.-O. A. Borrebæk, B. P. Jelle, and Z. Zhang, "Avoiding Snow and Ice Accretion on Building Integrated Photovoltaics – Challenges, Strategies, and Opportunities," *Solar Energy Materials and Solar Cells* 206 (2020): 110306, https://doi.org/10.1016/j.solmat.2019.110306
12. C. C. Ryerson and A. C. Ramsay, "Quantitative Ice Accretion Information from the Automated Surface Observing System," *Journal of Applied Meteorology and Climatology* 46, no. 9 (2007): 1423–1437, https://doi.org/10.1175/JAM2535.1
13. *Standard Test Method for Static Loading and Impact on Exterior Shading Device*, AAMA 514-16 (Schaumburg, IL: American Architectural Manufacturers Association, 2016).
14. L. Zabilansky, W. Burch, and T. Hall, "Laboratory Testing of Architectural Exterior Wall Fins," ERDC/CRREL TR-09-1, Cold Regions Research and Engineering Laboratory (January 2009), https://apps.dtic.mil/sti/pdfs/ADA536509.pdf
15. M. Carter and R. Stangl, "Increasing Problems of Falling Ice and Snow on Modern Tall Buildings," *CTBUH Journal*, Issue IV (2012): 24–28.